There'll Never Be Another
EWE

There'll Never Be Another EWE

WADE HUGHES
Illustrated by Don Lindsay

St George Books

Considering the relatively modest size of this book, there seems to be an inordinately large number of people who helped make it happen. Endless encouragement flowed from Robyn, my wife. Max Walker gave me a great boost. Brad Collis was there just when I needed him. Wayne Osborn stood out in the rain to take the author's photo. Jeff Knuckey painstakingly scrutinized the proofs. Julian Cribb and Ron Moon gave it their professional thumbs up. And way, way back in distant memory my Mum and Pop bought a ticket to Australia. To you all, thanks. — W.H.

Other books by Wade Hughes
KINGSFORD SMITH
DRUGS IN AUSTRALIA
POVERTY IN AUSTRALIA
EXPLORING AUSTRALIA BY SEA
EXPLORING AUSTRALIA BY LAND
EXPLORING ANCIENT AUSTRALIA
EXPLORING ANTARCTICA

First published in 1993 by St George Books — a division of West Australian Newspapers Ltd, 219 St Georges Terrace, Perth 6000
© Wade Hughes 1993
Fully set up on Macintosh computer by P.J. & T.L. Wells Typesetters.
Printed by Australian Print Group, Maryborough, Victoria. Cover by Pat Towers
All rights reserved. Except as provided by Australian copyright law, no part of this book may be reproduced without written permission from the publisher
National Library of Australia cataloguing-in-publication data:
Hughes, Wade, 1948- .
   There'll never be another ewe.
   ISBN 0 86778 050 9.
   1. Sheep-shearing – South Australia – Anecdotes. 2. Sheep-shearing – Western Australia – Anecdotes. I. Title.
636.3145

Foreword

There'll Never Be Another Ewe is a book that you won't want to put down.

It's a beautifully told tale of real life in the Australian bush, set against a backdrop of our shearing industry. Yet it's more than just a procession of good yarns. There is a lot of humour in this book. But there's a more sombre element too. Life in the bush is not always fun and games.

Wade spent seven years working in the shearing industry, absorbing the details that he's woven through this book. His warm and clear style illuminates his characters as though they're standing out in the brilliant sunlight of the Australian outback.

And I'm pleased to see that, in the best tradition of Australian yarn spinners, he's recognised that sometimes the facts need some adjustment to accommodate a good story. You can make up your own mind about Black Havoc — the world's most ferocious sheep dog; Kenny Dingle-Allen's mighty hit at the Kingoonya Cricket Ground; Pat "Bloody" Chiswick; Roodini, the escaping kangaroo; and the ghost of Billy Darby!

There was a time when people like those in this book were regarded as national heroes. "Australia rides on the sheep's back" was what they used to say. Woolgrowers who risked everything they had to create an industry, and shearers, the men who worked so hard to separate the sheep from the wool, were the pride of the country.

Tales of monumental shearing tallies were always enough to ensure that shearers, enjoying a drink in the pubs of our towns and cities, were fed with a never-ending supply of that cold amber fluid from a wide-eyed audience who couldn't believe what they were hearing — yet couldn't wait to hear more.

Of course, the old shearers were smart enough to realise that there's no point in letting the truth stand in the way of a good yarn — especially when you're thirsty. It's a philosophy I adhere to myself!

But times changed. Wool is no longer king. Like many other Australian export industries it's having to face up to some hostile, short-pitched bowling at the moment.

Australian society has changed, too.

Today it's a rich blend of people from many parts of the world.

Shearers and woolgrowers form just one small part of that society. And, because they bend their backs out in the backblocks, they're often forgotten.

There'll Never Be Another Ewe puts them back in the limelight. When you've finished it you'll realise that, not only have you been entertained, you've also had the chance to meet and get to know some of the characters who helped carve out the legends of the Australian bush.

And once you know them, you can't help but take your hat off to them.

MAX WALKER

Contents

Foreword — 5

Prelude — 8

CHAPTER ONE
Not a Good Start — 11

CHAPTER TWO
The Long Run — 33

CHAPTER THREE
Second Year: Second Prize — 51

CHAPTER FOUR
A Shiny New Stencil — 61

CHAPTER FIVE
Things That Count — 82

CHAPTER SIX
West of Centre — 92

CHAPTER SEVEN
Beach of a Life — 108

CHAPTER EIGHT
A Cool Change — 123

Epilogue — 136

Prelude

North Wales: mid-winter's night, 1958

Swirling snowstorm. Gusty sea wind blowing in over the wide estuary of the River Dee.

About a kilometre west of Flint, a small housing estate. Two-storey duplexes, huddling under the silent smother of thick snow, bathed in the sheen of moonlight-blue street lights.

Some people in bed; houses darkened. Others still up, swaddled in the hiss-popping comfort of coal fires. Their yellow lighted windows lonely outposts of warmth in the bleak landscape.

One by one the fires die down. The last smoke wreathes soundlessly out of chimneys and disappears into the snowflecked darkness. The yellow windows go dark.

All except one. Upstairs in the house on the corner, behind the checkered curtains, a little boy reading in bed. A comic. *The Wizard*. Not a picture comic. More like a feature newspaper. Long stories. Serious stuff for a nine-year-old.

Tonight's story is about the Australian outback. The little boy couldn't put it down until he finished reading it . . .

Blonnng! Blonnng! Blonnng! Soft muffled chimes from the mantle-clock downstairs. End of the story. Time for sleep.

Bare feet on cold lino. Across to the light switch. The last outpost of yellow surrenders to cold blue. Back into bed. Burrow deep under the covers. Cold hands jammed between pyjama-warm thighs.

Wind-rattled window panes. Distant shriek of the Holyhead

Express storming down the rails on its way to the coast. Stirring thoughts of adventures in the outback. Cattle duffers. Bushrangers. Black trackers. Bounding kangaroos. Stampeding wombats. Deep sleep . . .

Cocklebiddy, Western Australia: mid-summer, 1968 6.25 A.M.

Bratatatatatadang! Iron bar rattling against corrugated iron. "Righto, you bastards, breakfast's ready. Come and get it!"
Suddenly awake! Sweat-damp sheets. Bare feet on warm concrete. Squint through the glassless window. Bright light, hard on the eyes. Stunted pale-blue saltbush. Rust-brown earth. One hundred metres away, grey iron shearing shed. Hot, woolly sheep pressed into cyclone-mesh yards. Smell of urine-damp dust. Fanfare of the ABC news. Cultured headlines. Barking dog.
About two metres away, Pat "Bloody" Chiswick belched, broke wind, sat up in his bed and started a lung-wrenching coughing spasm which culminated in an heroic attempt to propel an olive-green lump of phlegm through the open door. It fell short and landed like an exhausted jellyfish on the doorstep. "Bloody hell, what a dump, what time is it?"
"Six-thirty."
"All-bloody-ready?"
New sights. New sounds. New smells. The little boy had grown up. It was my first morning in an Australian shearing shed.

Portland, Victoria: 1993

I went into the sharp end of the Australian wool industry — into the shearing sheds and woolstores — because I wanted to see, and become a part of, the Australian outback. Not because I particularly wanted to be a woolclasser.
There's a rich culture out there in the sheds; it is probably the last real preserve of old colonial Australian bush life. Miners and

farmers have all moved on into varying degrees of high technology. Ore bodies are found with the help of satellites. Lasers help level paddocks.

But shearers work and live in much the same way they have always done. Their working life is an anachronism and it's in danger of extinction. Shearing robots are being given life in workshops and laboratories around the country. High-tech instruments are measuring and assessing wool much more accurately than human hands and eyes could ever hope to.

Perhaps robots and computers will never be able to replace manual labour in shearing sheds, but then, people thought that Lawrence Hargrave's experiments with flying machines were a waste of time, too.

So, now distant from the industry I learned to love, I decided to compile this chronicle of life in the bush as I saw it.

It's simply a record of the hard work, the humour, and in some cases the bitterness that makes up life in this most distinctive of Australian industries. It is my attempt to record the characters and events before time and changing Australian values bury them for ever.

All the incidents are based on fact, but some are truer than others. I've changed the names of some of the locations and of some of the characters, too. If those blokes had wanted publicity, they wouldn't have chosen a career in the bush.

Oh! There's one more thing. Seven years in the bush also taught me that whoever wrote those articles in *The Wizard* had never been to Australia!

WADE HUGHES

1

Not a Good Start

"A WOOLCLASSER! NOT as long as I live and breathe!" That was Pop.

"You're going to uni and that's final!" That was Mum.

They always saw things in clear-cut black and white. They'd sacrificed a comfortable life in Wales to give their kids a fair go in a new land. They predicted that the little factory town they lived in could offer their kids very little. (They were right. The factories have gone and rank unemployment now stalks the streets.)

So, in their forties, they abandoned the life they knew, severed their own family ties, and with about two bob in their pockets, sailed off to God-knew-what in Australia.

Thousands of other British families were doing the same thing; and hundreds of them were returning to the Mother Country beaten and disgruntled. Whingeing Poms.

But my sister and I were told: "We will not be going back."

We all boarded the train at Flint Station in North Wales. The guard blew his whistle and the door slammed on our past. Six weeks later we arrived in Adelaide. I was ten years old.

Eventually we settled in a new subdivision at Christies Beach, south of Adelaide. It was surrounded by farmland, and by being forward I got to know Clem Dyson, one of the biggest local farmers. Clem let me think I was working on his farm during my school holidays and he taught me the rudiments of sheep work.

On one particularly gory evening, as the red disk of the sun sank into St Vincent's Gulf and dusky calm settled over the paddocks, he let me slaughter a sick wether. He made me go back to it every ten minutes after tea to see if the operation had been a success.

After the third visit I was able to report that the unfortunate beast's struggles had finally ended and I was able to carve the first notch in the handle of my Green River slaughtering knife.

Part-time life on the farm made me certain that I wanted to work on the land.

High school years slipped by amidst boisterous classmates wrestling their way through hazy days of obscure algebraic formulae, English literature, clumsy chemical experiments in the laboratory, and clumsier biological experiments in the tiny dressing room behind the stage.

But suddenly it was over. The day had arrived when we had to make a decision.

"What are you going to do when you leave school, Wade?"

"Er . . . um . . . take up woolclassing!"

That's when my parents made their feelings known. Tempestuous arguments followed, but in the end Pop took me to town to enrol at the South Australian Institute of Technology.

There were more applicants for the woolclassing course than available positions, but I knew I must have had a fair chance of acceptance when I overheard a brief conversation between one hopeful lad and the senior lecturer. The lecturer was looking at the lad's application form.

"What's your name, son?"

"Cyril, sir."

"That's funny, my name's Cyril too — but I don't spell it like that."

"Oh, er . . . um. Well, I might have spelled it wrong then . . ."

Two weeks later the institute wrote and said I'd been accepted. That was beaut. But I still had to earn a living.

My parents couldn't afford to support me while I attended wool school full-time. So I enrolled as a part-time student and went job hunting.

Tall, boxy, red-brick woolstores at Port Adelaide were the best bet for an aspiring woolclasser. I tried Dalgety's first and they gave me a job.

Thankless, dirty work, carting wool in wheeled baskets from the bulk-classing and reclassing floors to the type bins.

This was wool which had either been poorly classed or not classed at all in the sheds. Monotonous though it was, it gave me an introduction to wool and the men who worked with it. They all urged me to find another career!

Six months went by. Long days in the stores. Night school twice a week. One night of theory at the institute, the other doing practical classing on handkerchief-sized scraps of wool in the wool pavilion at the Wayville Showgrounds.

We learned what woolclassers look for in fleece wools. For merinos, the length of the clumps of fibres — they're called staples — is important. Generally about ten centimetres for first grade, shorter for lower grades.

The size of the waves (or crimps) in the fibres told a story, too. The more crimps to the centimetre — the finer the wool. The finer the wool, the more hanks of yarn that could be spun from it. Strength had to be judged also and colour was important. Good wool is pearly white. Yellowing reduces the value.

Yield was paramount. That was how much usable wool could be extracted from a kilogram of wool straight from the sheep's back. All sorts of things affected yield; the amount of woolgrease, or lanolin, in the wool, the amount of sand or dust, and the presence of small sticks or burrs.

We learned to class belly wool. Bellies are the first wool to come off the sheep. They're taken off in a single piece about as big as a table napkin. We were told about pieces, the rough and sweaty edges that are ripped from the edges of shorn fleeces. And we learned about locks. Tiny snippets of wool, cut from the sheep when the shearer doesn't get some of the wool off cleanly on his first attempt.

Crossbred sheep were another matter. Wool like rope, up to thirty centimetres long. Crimps like the corrugations in corrugated iron.

Then Dalgety's made me their wool cadet. They'd pay all my school fees and a full adult wage. My parents were ecstatic. Mum got straight on the phone and told her mates. Pop shook my hand and poured us a beer.

Night school wasn't going well, though. Wal McGovern, one of the lecturers, worked at Dalgety's and he took me aside one day to offer me some extra coaching. When I said I didn't need it, he offered me last year's exam paper.

"See how you go with that, then."

Dismal failure.

That did it! There were three weeks to go till the final exams and I could just imagine what my parents would do if I failed.

So I gave up all the pleasures of the flesh and worked hard, driven by fear more than a desire to learn. Every night. Every lunchtime. Poring over books. Scribbling notes. Desperately trying to form a clear idea of the various wool types.

Three weeks is not a lot of time to make up for a whole year's bludging and I had some misgivings when I handed in my paper. But it all came good. When the results were published Mum got to the newspaper first. With tears in her eyes she rose from the table and hugged me. I'd passed with a distinction — the only one awarded. I'd topped the State in the first year course.

"Struth!" You could have knocked me over with a lamb's tail!

☆ ☆ ☆

But you couldn't learn to be a woolclasser just by going to night school and working in the woolstores.

Trainee woolclassers had to work their way up through the ranks in shearing sheds known as school sheds.

These were ordinary contract sheds but the woolroom was staffed by trainee woolclassers, or "schoolies". They started as lowly rouseabouts, and over three years worked their way up to be foremen — a sort of woolclasser's right-hand man.

Then they were granted a probationary stencil and turned loose on the Australian woolclip.

Woolclassers' stencils are their signature, branded on every bale they class. When the probationary classers felt confident enough they had to submit two clips for appraisal. Then, if they'd learnt their trade, they were awarded a full professional stencil.

So now, with a year's technical school theory out of the way, I was ready to be sent to the bush to start my apprenticeship in the sheds. The adventure was about to begin.

WEAR OUT A BROOM A WEEK

"YOU WIDE EWES?"
   Pop and I were on the old Adelaide railway station in North Terrace. I was waiting to board a train, my destination a shed over the border in Western Australia.
   The bloke behind the strong Australian accent had piercing blue eyes, short back and sides, baggy trousers and two dogs constantly growling, circling each other and knotting their leads round his legs. He was John Murphy, the shearing contractor.
   "Yes, that's me. You're Mr Murph . . ."
   "Call me John. 'Ere's your tickets, foller me." He turned, kicked the leads off from round his legs and shuffled away, dragging his scruffy dogs behind him.
   Pop and I had time for a quick handshake and I was off, following Murphy, dragging my shiny new suitcase on its wheels behind me. I had a sudden, frightening thought.
   "Bloody hell! I'm leaving home!" I handed the ticket to the porter and stepped on board with the rest of the new team.
   Jigsaw puzzle fragments of my surroundings crowded in on my senses and piled up in a corner of my memory. Crowded compartment. Smell of sweat. Glimpse of Pop through the window, looking for me. A tap on the glass but no chance of him hearing: double glazing. Sudden tears welling up. Distant whistle. "All aboard, please!" Slamming doors. Floor trembling underfoot as eighteen hundred horsepower takes up the strain and almost imperceptibly eases the train forward. Slow-motion parade of passing faces. Waving. No sign of Pop. Hard seat. Smell of vinyl.
   Soon the train was rickety-racketing through Adelaide's northern suburbs. Gloomy backyards. Tumbledown fences. Pigeon roosts. Clothes hoists full of limp washing. Kids staring at the train. One swung his arm around . . . the little bastard threw a rock at us! A faint reflection of myself in the window glass caught my attention. Eyes still red and staring at me. "Too late now, mate . . ."
   Before I'd left the woolstores one of the classers had given me some advice. "If you want to do well, keep the shed tidy for the

woolclasser. Wear out a broom a week." I wondered how many brooms I'd be able to wear out on this trip.

After sitting up all the way to Port Pirie it promised to be a pleasure to transfer to the comfort of the East-West Express for the main leg of the trip through South Australia and over the border across the Nullarbor Plain. At last, into the bush! We disembarked and trooped in a rag-tag throng across the Port Pirie station platform and boarded the express.

More jigsaw puzzle memories. Still, chill, air-conditioned atmosphere. Congested, serpentine passageway. Tiny compartment. Two seats, facing each other. Flip-up table hinged to the wall between the seats. Flip-down wash basin behind the door. Drop-down bunk with a stainless steel locking mechanism.

Then a tall, rough man squeezed himself and his bundle of baggage from the crowd in the passageway. The tiny compartment was full.

Faded denim dungarees. Frayed tartan flannel shirt. Battered suitcase. Carton of West End cans and a scuffed leather kitbag. He mumbled a half-hearted introduction: "Pat 'Bloody' Chiswick."

That was some train ride. Pat Chiswick had drunk half his cans almost before we'd cleared the outskirts of Port Pirie. He was legless by the time we rattled into the night somewhere out to the north-west of Port Augusta. So were most of the rest of the team. But that didn't stop them from staggering from compartment to compartment in a sort of progressive party.

There were five drunks and me in our compartment at one stage. Then they suddenly decided that "Robby" should be visited. Off they went singing and yelling. *"Robby! Robby! Robby ya bastard. Open the bloody door!"* After Robby it was Klaggy. *"Klaggy! Klaggy...!"* They visited everyone's compartment in turn. In their wake they left a filthy avalanche of cigarette butts and empty cans.

Other passengers complained. The conductor reacted swiftly but feebly. *"Now look, you blokes..."* Someone offered to stuff him down the dunny and he retreated to his own compartment. Sometime just before dawn the noise finally died down.

I went to the first sitting of breakfast and walked into a tense, hostile atmosphere. The other passengers were furious and very tired. Very few of the team showed themselves for breakfast. They even skipped lunch. Muttered apologies from those who did show did nothing to help ease the ire of the passengers. It was a long day.

About four in the afternoon we pulled into Rawlinna, a dusty, scorched railway town shimmering on the Nullarbor Plain. This was where we got off.

As we filed down the passageway, I glanced into one of the compartments. One of our blokes had left a parting gift for the conductor. He'd defaecated on the little swing-up table that sits between the seats.

Outside a heap of baggage grew on the station. Murphy's dogs appeared from the baggage car. A whistle blew. A door slammed. The motors throbbed more urgently and the train pulled away, scorning us with angry stares from accelerating white faces at the windows.

"Jeez! She's a bit bloody tropical out here!"

"Where's the cocky, Murph? 'E's s'posed to pick us up, i'nt 'e?"

"'E'll be 'long d'rectly."

The cocky was late. A long, stifling afternoon droned into suffocatingly hot twilight. We sat uncomfortably on the Rawlinna railway station. Most of the team topped up their drunken spirits and sang the time away.

Just on dark a small, open wagon trundled past on the railway track. On board was what looked like a bunch of pirates. Rough, hard men, tattooed, smoking, swearing. Two of them were pumping the seesaw lever that propelled their wagon. Railway fettlers. They'd been working on the line way out on the plain. As their wagon squeaked to a stop past the end of the platform, Klaggy voiced his thoughts. "I s'pose if yer'd been working out there all day, yer'd be glad to get ter Rawlinna. Can't see any other reason fer liking this place meself."

Twilight deepened into purple night. More stars blazed overhead than I'd imagined ever existed.

Six hours after we'd arrived the cocky turned up with our transport to the shed, 100 kilometres out across the plain.

"Pile in!"

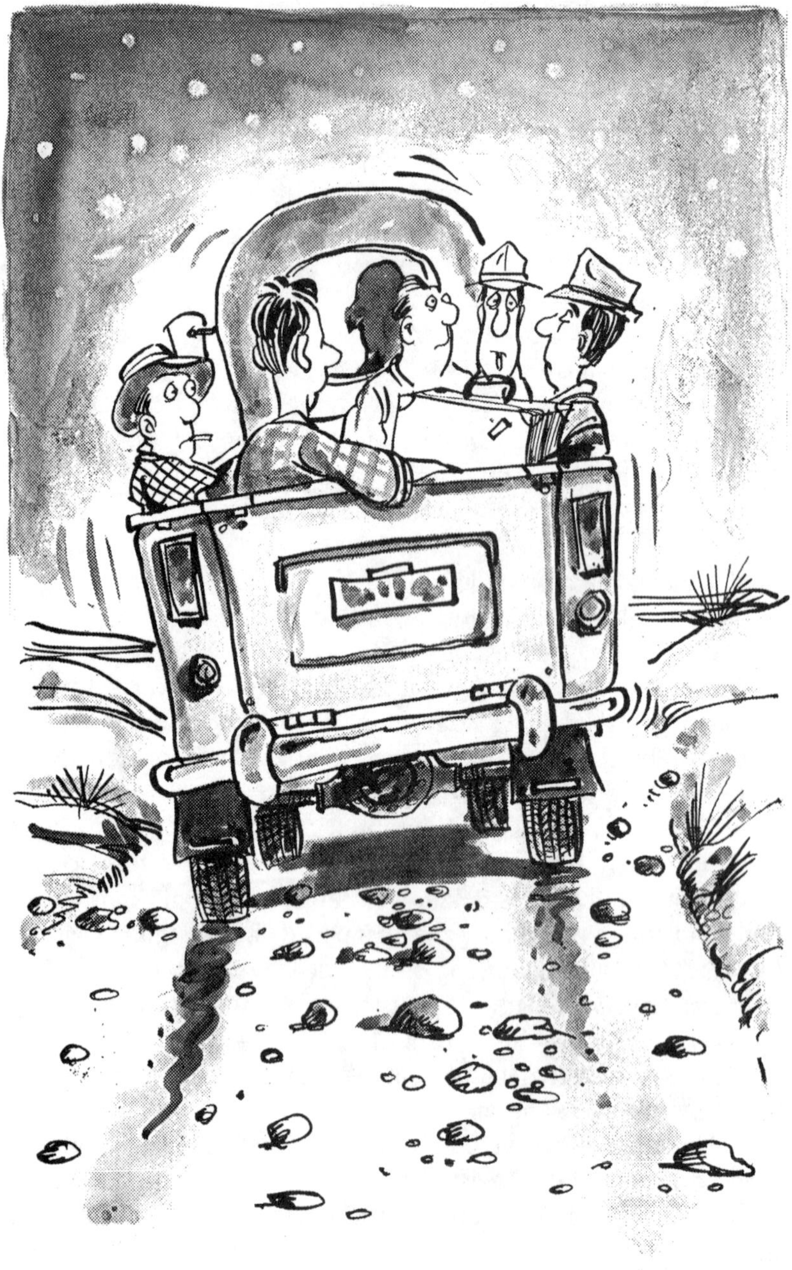

One truck. Two utes. Suitcases. Men. Dogs. Stores. Rolls of bedding. Cartons of soap powder. Boxes of beer. All heaped in untidy jumbles on to the vehicles.

I ended up on the back of a ute with Jerry, another rouseabout. I was the only schoolie. He was the board rouseabout. He was going to be a shearer.

"Everybody right? Hold on." Roaring motors. Lurching bodies. Yellow beams of light scything across the flat as the vehicles turned out of the station yard and headed off down the track.

Startled rabbits bolted, then propped in the lights. Long, black shadows speared away from every roadside bush. Black bars banded the track ahead. "Corrugations," said Jerry. Corrugations that pounded my bum on the hard wheel-arch of the ute. Corrugations that made my back ache. Then the taste of dust. We were the last vehicle in line, so we copped our own dust and everyone else's. Powder-fine. Swirling clouds of it. Clogging the nostrils. Gritty between the teeth. Irritating the eyes. Now I knew why everyone else in the team had boarded the train in scruffy working clothes.

It was a long drive. The blokes in the front of the ute didn't mind, though. They were travelling in relative comfort — and they still had a few beers left. Every half hour or so the passenger window came down and an arm sent a brown bottle or dead can spinning away into the bush. Nobody offered Jerry or me a drink. Not even when the three blokes in the front had to stop to drain their bladders. They didn't waste the stop. Robby rummaged through the dust-covered boxes in the back of the ute and found a carton of warm beer.

Three hours after we'd left the station we arrived at the shearers' quarters. "Thank God for that!"

A 32-volt lighting motor thumped away somewhere back of the quarters. Washed-out yellow light seeped from the curtainless, glassless windows and formed thin pools on the ground outside. This was home for the next three weeks.

Somehow the dishevelled loads of gear were sorted out into recognisable heaps. Suitcases. Men. Dogs. Stores. Rolls of bedding. Cartons of soap powder. Boxes of beer. Then the utes and truck revved off in the direction of the homestead and we were left alone to find our rooms and make our beds.

I took the closest room I could find. There were two beds: cyclone mesh wire strung over a steel frame and bare mattresses and pillows. Pat "Bloody" Chiswick followed me in. Good and drunk, he didn't bother with a sheet.

When he'd precariously wobbled his way out of his shoes and trousers, he just heaved a thick blanket from his bedroll, performed a sort of tottery pirouette to wrap himself in it and crashed on to the bed farthest from the door. Within moments he was snoring. I fussed with sheets, pillowcases, blanket and dustcover. At last, time for sleep. Pull the light cord. The last light to go out. The automatic lighting motor cut out and died away.

☆ ☆ ☆

Bratatatatatadang! An iron bar rattled against corrugated iron walls. *"Righto, you bastards, breakfast's ready. Come and get it!"*

I sat up in bed and squinted out through the window. The outback!

Then the sound of two dogs fighting, behind me, in our room! My head snapped round to see the action, but it was just Pat "Bloody" Chiswick coughing.

He wiped his mouth with the back of a hairy hand, threw off his stained blanket, and swung his bare legs out of bed and on to the floor.

Pat was one of those shearers who had gone bush as a lad and spent his whole working life beyond the fringe of civilisation. His legs were only slightly thicker than those of a freshly shorn sheep, and about the same colour. When they hit the floor it seemed unlikely they would support the weight of his ponderous belly. But he took the risk of them collapsing and, flinging his towel over his shoulder, lurched out the door, furiously scratching his bottom and looking more like a refugee from an old men's home than a gun shearer.

Elsewhere in the camp a few radios sprang to life and there was a general sense of movement at the station.

I got out of bed and opened my new but instantly aged suitcase.

Neatly pressed and folded khaki drill trousers and shirts were still that way but already looked as though they had done a tour of duty in the bush. Thick dust had burrowed its way into my suitcase.

"Yer've got to wrap everything up in plastic bags," Pat "Bloody" Chiswick told me as he came back from the shower block.

Woolclassers in the stores wore khaki drill trousers and shirts. That's what I thought woolclassers and trainee woolclassers would wear in the sheds. But when I walked into the shower block I realised that no one actually dresses to work in a shearing shed. Blue singlets, tatty dungarees and grubby old windcheaters formed the uniform of the rest of the team.

Trevor, another of the shearers, looked up from the threadbare depths of a well-worn towel. "G'day Blue! Ready to raise a sweat, are yers?" Everybody with red hair is "Blue" out in the bush.

Outside it was already warm. The messroom was stifling, with its blazing wood stove and sweating cook. Breakfast was bacon, eggs and liver. I bolted it down, filled my brand-new canvas waterbag with rainwater from the tank outside the mess and headed for the shed, leaving a trail of water in the dust behind me.

"Yer've got ter soak them new bags fer twenty-four hours before they'll 'old any water, Blue," Murphy said on his way to the shed.

I stopped to take a look around, much as a tourist does when he first sets foot in an exotic location. This, after all, was *the* Nullarbor Plain. The bush. The outback.

Murphy's dogs loped past, seemingly oblivious to the heat.

Bingle-Bingle was a relatively new shed. Steel frame covered in corrugated iron. It looked like the kind of shed you'd expect to find on the shelf in an Elders store. That's probably where it came from. Inside everything was neat and tidy. There wasn't quite the atmosphere or romance that I'd expected.

The shearers straggled in. The classer came in and introduced himself: Syd. Fat, red-nosed and red-cheeked. He'd driven here from his last shed on the west coast of South Australia.

Murphy heaved his massive gut up the steps from the engine room and placed a sheaf of papers on to the wool table. These were the shearing agreements, signed before the start of shearing at every shed. The whole team gathered round. Four shearers, woolpresser/penner-up, board rouseabout, and me, woolroller, working on the classing table with the classer, helping him to rip the rough edges and other poor quality wool from the fleeces.

At half past seven, Murphy belted one of the shed's steel uprights with a hammer and said: "Righto you blokes, into it!"

Four shearers strode into the pens, caught their first sheep and dragged them out on to the board. Shearing at Bingle-Bingle had begun.

All the shearers finished their first sheep within seconds of each other. Gleaming white sheep, short-back-and-sided all over, clattered and scrabbled down the let-out chutes into the count-out pens. The shearers strode across the shorn fleeces, into the pens, to catch the next sheep. Jerry sprinted up and down the board, frantically grabbing the fleeces. He threw the first one out like a blanket on to the slatted wooltable and dumped the next three on to the floor at the foot of the table.

Woolrolling seemed simple enough. But it wasn't. Everything I did seemed to be wrong and Syd quickly became irate. The next fleeces arrived and we still had two of the first four on the floor. Then there were eight on the floor. Then eleven. Sweat started to run freely. My fingers, knuckles and wrists all started to ache from the constant, high-speed grabbing and tearing. Dust filled the air. My throat was parched. Finally, in the middle of the frenzy, I just had to have a drink. Two greedy gulps drained my leaky bag — but at least the water was refreshingly cold. Then my eyes settled on the board clock. Only ten to eight! It was going to be a long day!

"Grab the bloody broom!" Syd shouted.

I'd forgotten all about brooms. The floor was covered in a snowstorm of tiny snippets of wool, just like a barber shop's floor. These are called second-cuts, because they're the result of the shearers taking two cuts to get them off the sheep. The first cut often leaves ridges of longish wool still on the sheep. Second cuts get rid of them. Wool growers don't like to see them because they're a waste of good, full-length wool. Woolclassers don't like to see them because they get mixed up with the longer fleece wools and the woolbuyers complain. So the idea is to keep the floor well swept of these nuisance second-cuts. They're pressed into their own bales and branded LOCKS.

I grabbed a straw broom and feverishly thrashed the floor with it. "More speed! More speed,"yelled Syd.

There was a half-hour break at nine-thirty. The cook brought hot scones and scalding tea from the mess. It tasted like the Last Supper to me. At midday we stopped for lunch and trooped back to the mess. A grimy thermometer hung by the door on the verandah. It showed 121° F.

This was all a bit of a shock! By the time afternoon smoko came at three o'clock I knew exactly what I was going to do. Knock-off time was five-thirty.

After tea I flopped down on my bed and somehow rallied enough strength to grab my writing-pad and scratch out a letter.

"What, writing a bloody letter home all-bloody-ready!" Pat "Bloody" Chiswick exclaimed. "Bloody homesick rousea-bloody-bouts." He knocked the top off another bottle of beer and settled down to a *Phantom* comic that someone had left in our room. I turned back to my furtive scribbling.

Dear Mum and Pop, well you were right. Woolclassing is not what I thought it would be. I'll come home straight away and probably go to uni or something. There is one problem, though. I've had to sign a shearing agreement and I can't just walk out of the shed without good reason. John Murphy won't pay my fare back home if I do. I haven't got enough money to pay my own fare, but if you send me a telegram saying Mum's crook, I'll be able to get out. All my love, Wade.

That letter had to wait two days before I could give it to the owner's son, who was going into Cocklebiddy to get stores. He posted it for me and my spirits lifted. Not much, though.

The days dragged by painfully. My knuckles and wrists were in agony. First there was "woolroller's wrist". Constant grabbing and tearing to remove the straggly edges from the fleeces made my wrists swell and ache abominably. Everybody gets it. Today it's probably diagnosed as RSI. The inflammation and swelling spread down to my knuckles. Overnight my hands would contract like hooked talons and set solid. Freeing them in the morning was painful enough to bring tears to my eyes.

In the end I had to soak my hands in hot water for ten minutes or so before I could bear the pain of trying to straighten the tortured tendons. During the day the slightest knock on any of my knuckles brought shooting pains through my whole hand.

Then there were the prickles. Wool is often liberally spiked with grass seeds and prickles, which frequently puncture soft fingers. When they go deep they feel like electric shocks and they have to be dug out with a needle or the sharp end of a bale fastener. Shearers use the points of their cutters to dig them out. The most painful prickles are the ones that go deep into the back of the finger, just below the nail. Sometimes a prickle would go in too deep to be dug out. The only thing to do was to let nature take its course and wait for the wound to fester. Then, when a throbbing green head appeared over the prickle hole, it was relatively simple, though painful, to squeeze it and watch the offending prickle pop out.

So with RSI and prickles my hands were in a sorry state. Gripping anything was an unpleasant experience — and that included a knife and fork. Eating became such a chore I felt in danger of starving to death.

Night shift was a problem, too.

Nobody shears sheep at night but young tortured woolrollers often skirt fleeces nocturnally. I knew when I'd been doing it. I'd wake up feeling more tired than when I'd collapsed into bed and my bedclothes would be bundled up in a heap on the floor all neatly woolrolled!

On top of all this the classer kept shouting at me. And this was only the beginning.

After three weeks at Bingle-Bingle we were due for a three-day break in Adelaide before driving north to the Flinders Ranges to complete another two sheds.

Normally a first-year woolclassing cadet is only required to suffer four weeks in the sheds: for some reason I'd been given a longer run, nearly ten weeks "to let him develop".

Develop? I looked around me. Four shearers, bent double, straining and cursing as they drove themselves harder and harder to reach their tallies. Tight-pressed sheep standing in the pens, constantly grinding their teeth. A board rouseabout running up the board with arms full of shorn wool, then running back again to sweep the board clean before the shearer returned with another sheep. The woolpresser's veins bulging in his face and neck as he squeezed every last ounce of his strength into pumping the press handle. The furrowed brow of the woolclasser as he mentally wrestled with woolly decisions, constantly scurrying about the woolroom checking and double-checking his bins, then returning to the wooltable to rip the edges from the next fleece, all the time shouting at me, "More speed, more speed!"

That was inside the shed. It was even worse outside. We were 600 kilometres over the South Australian border into Western Australia. Outside lay a landscape so flat and featureless that by standing on the roof of the shed it seemed almost possible to spot Sydney Harbour Bridge!

Develop? Vegetate was the more likely result.

Then the telegram arrived! With embarrassing eagerness I leapt forward as soon as the cocky walked into the shed with it in his hand. He'd just taken it down over the radio-telephone in the homestead. "Hope it's not bad news," someone offered as I unfolded that priceless piece of paper. I tried to look concerned but my heart was leaping with elation.

Damn! My jaw dropped. It *was* bad news. The kind that prompts suicide. *HOPE YOUR BED IS COMFORTABLE. REGARDS MUM AND POP.*

Most of my life my parents had been telling me that "If you make the bed, you have to lie on it". The cocky looked at Murphy and grinned. I sulked for two days.

GETTING TO THE BOTTOM OF IT

TIME GROUND ON. Interminably. I felt trapped in a hostile environment. Homesick. And isolated from the rest of the team because I was so new and green. And useless. And a schoolie.

Murphy threatened to set fire to the school when he got back to Adelaide — revenge for sending "another useless bloody layabout". Syd had it in for me, too. I hadn't picked up the knack of speedy woolrolling. "You'll have to get more speed. More speed!"

Nearly every morning I'd wake up to find I'd woolrolled my blankets in my sleep. Then Klaggy started threatening to jump into bed with me and "pin a tail on the young 'un".

Klaggy was a small, wiry shearer. He wasn't a homosexual, he was just practising a time-honoured ritual of initiation. In hindsight, it's harmless and at the end of it all, if the initiate survives, there comes acceptance, respect, even affection, from the rest of the team.

But that's with hindsight, and Klaggy went too far. One boozy Saturday night he left the party in the mess to drain his bladder. That's when he peeped in through my window and found me lying in bed reading.

"I'll be in there d'rectly, young 'un!"

"Piss off!"

"Oh! Don't get nasty now!" And he scuttled back to the mess —

but only to raise some support.

I could hear snatches of his plan over the general clamour of the party.

"I'll give 'im a lovebite to go 'ome with . . . that'll give 'im somethin' ter explain ter 'is mother . . .!"

Two or three of the blokes came back with him. "Pucker up, young 'un. 'Ere I come!"

He piled on top of me — but he was drunk and I was sober. I wriggled out from under him and left him in a giggling heap in my bed. "Come on, young 'un. Don't play hard to get!"

Everybody thought it was hilarious. Except me. All the frustration, all the despair came to the surface. My temper snapped. I grabbed the foot of the bed and heaved it up, tipping Klaggy down towards the head. When I rammed the bed over and into the wall, he was upside-down and squashed. When I let the bed go, he fell in a heap on to the floor, head first.

Suddenly Klaggy didn't want to make love to me. He wanted to kill me! Then Murphy stopped things from getting any worse and the party in the mess soon got going again. I remade my bed, climbed back in, covered my head with the blankets and cried as quietly as I could.

☆ ☆ ☆

Cut-out came at last. We washed down the shed, swept out the quarters and clambered into the utes and truck for our ride back to the East-West railway line at Rawlinna.

Rawlinna hadn't changed much in the three weeks we'd been at Bingle-Bingle. Rawlinna probably won't ever change. Around the town the treeless plain sits baking in the heat. It used to be a seabed, laid down in the cool depths when the western half of the continent was under water. Now only heatwaves wash across its surface. You'd need to be a special sort of person to want to live in Rawlinna. We just wanted to leave — but we couldn't. We'd been banned from travelling on the East-West Express!

Murphy was furious. The stationmaster didn't even apologise for the inconvenience. "Yer own bloody faults." He did call Kalgoorlie, though, and told them we'd arrived. It had all been arranged. "They'll hook a van on to the back of the next fast freight to leave Kal."

"How long before it gets 'ere?"

"Eight hours or so." An hour later the East-West Express rocketed through without stopping.

"It's because of that turd you left em, Klaggy."

"Don't be stupid," responded Klaggy, "how could they know it was me!"

Time passed very slowly. A brief diversion occurred when one of the locals opened fire with a pump-action shotgun. His target was a wild turkey which had been silly enough to land in the siding. These turkeys — really bustards — are protected by law because of their rarity, but that doesn't seem to stop anyone who spots one from taking a shot.

This one was more fortunate than most. Its would-be killer was drunk, so the barrage of buckshot went wide of the mark. It tucked its long, thick legs away and took off into the sky. Silence, save for the buzzing of blowflies, settled on Rawlinna again and we all retreated into our private thoughts. It was too hot to talk. Time crawled by.

There was a grey concrete toilet block at one end of the railway station. It fitted well into the surroundings. John Murphy stirred. "Anyone in the dunny?"

"Na."

He struggled to his feet, arched his aching back, hitched his pants up so the baggy crutch rose to just above his knees, belched, then shuffled off towards the toilet block.

About the same time I felt the need to use the toilet, too. Murphy had disappeared into the bowels of the building by the time I got there so I wandered around outside waiting for the grunting from within to stop. Then a trapdoor set into the rear of the building caught my eye. I'd never seen a thunderbox with a trapdoor before, so I squatted down and opened it.

Of course it contained the five-gallon drum into which the effluent fell. Although I almost recoiled in disgust I noticed that, directly above the drum, as you would expect, was the seat and on that seat, visible through the hole, was Murphy's alabaster bottom. I poked that bottom with a handy stick.

The results were immediate. Murphy let out a howl of horror. "JEEZ!!! What was that!" Only someone who has apprehensively lowered their naked and most sensitive anatomy on to a cobweb-swaddled thunderbox seat can really appreciate what must have gone through his mind.

The other blokes thought it was funnier than Murphy did. It helped break the ice. After three weeks, the blokes in the team finally laughed with me instead of at me, and I felt like a human being again.

Eventually the freight train arrived and then we all felt like cattle. Instead of a pleasant ride in air-conditioned comfort, we were to be shipped back in a prewar wooden carriage. No seats. Bare wooden bunks. No mattresses. No fridge. No bar. Just an old stainless steel urn in a little cubicle at the end of the carriage. Great for making plenty of cups of tea. The only thing was there was no tea on board. No sugar either. No food at all.

"'Ow long we got before we leave?"

"Ten minutes."

Robby and Jerry raced off to see if they could rouse the storekeeper, while the rest of us rummaged through the boxes of Murphy's unopened stores left over from our stay at Bingle-Bingle.

Murphy itemised the results. "Two tins of Monbulk melon and lemon jam; one tin of Milo; one tin Keen's mustard; one big tin Sunshine powdered milk; six packets Bushells tea; half a bottle of Chicko coffee essence; eight bottles of Heinz tomato sauce; one bottle of Worcstsrer . . . Woortssrchter . . . Wootstshooisher . . . Holbrook's bloody sauce!"

And that was it. Fortunately the storekeeper was a humanitarian and he whizzed around to open up the shop. Robby and Jerry cleaned him out of tins of corned beef, bread rolls and camp pie. And a carton of Swan lager cans.

"Did yers get some sugar?"

"Na! Forgot all about it. Got some tomato sauce, though."

We made ourselves as comfortable as we could and the train lurched out of the station into the night. Back in the 1940s, our carriage had been used to carry troops. I could tell: *Reg Doonie. 148687. 2nd AIF 1942* was carved into the wooden ceiling over my bunk.

It was a long trip. Jolting stops woke us frequently during the night. Voices in the dark: "What's going on?"

"Dunno."

Ten minutes later, the answer: a thundering roar. Speeding windows of light rattling past.

"The express. 'Eading for Perth. We've pulled over to let it past."

We pulled over to let everything past! Twenty-seven hours of misery finally ended when we rumbled into Adelaide.

I really was beginning to think I'd made a mistake about opting for life in the bush. But things were about to get better.

SPEED AT LAST

HOME NEVER FELT SO good. The beach. The clear water. The peace. No sheep. No woolrolling. No Syd. He wasn't coming on to the next sheds up in the Flinders Ranges.

Gill Braundfeldt was, though. Another schoolie, he would be foreman for the next two sheds. Gill was a good bloke who'd worked in Dalgety's with me. At least, a familiar face and a kindred spirit. The classer was to be old Henry Knight. Fussy, but good fun. A strict disciplinarian, but fair. It promised to be a better time all round.

Even so, the three days' leave went far too quickly. Gill and I were soon on the road together, heading for Bertaratna, a picturesque little shed in the heart of the Flinders Ranges, beneath towering peaks.

It had corrugated iron walls and a flagstone floor. It was dark. Too dark for Henry Knight early in the morning. "Can't see a thing!" he exclaimed. He stood outside and admired the red and purple scenery until the light improved.

Meanwhile, Gill and I kept rolling fleeces and stacking them on and around the classer's table. It didn't take Gill long to realise how slow I was. It became obvious to him by the time we had fourteen fleeces lying at the bottom of the table. "Come on, Wade. Pull your frigging finger out!"

"I'm trying. I am trying."

Then Henry walked in. And exploded. "Get that wool cleaned up and do it now! Or you'll be out on the next mail truck!"

Gill slipped into overdrive. He was skirting his side of the fleece as well as half mine — and we were still getting snowed under. My hands just couldn't move any quicker. Henry stood and watched for about five minutes. Fleeces kept building up at the foot of the table. Then he moved in.

"Stand back. Now watch me and Gill!"

Embarrassed again, I watched carefully.

"See?" Deft flicks. Straightening the fleece out. Nimble fingers picking instead of great fists tearing. "It's easy." Economy of movement. "You've got to keep at it — but you should be able to keep up. And find time to keep sweeping the floor. Wear out a broom a week and I'll make sure you pass. Okay?" A friendly slap on the shoulder. "Give it a go!"

Gill offered advice as well and Henry kept encouraging me — with only a mild rebuke when he found a great dag in the middle of a prime fleece.

By the end of the shed I was keeping up and enjoying it. At last I'd found speed. RSI had come and gone. Even Murphy had to admit I'd improved and the school was safe from arson.

When we walked out of Bertaratna my red-handled broom was leaning exhausted in the corner. Worn out.

☆ ☆ ☆

Our last shed was fifty kilometres farther up the track. Yantaboo, near Mount Searle. That was like a holiday camp. Good sheep. Easy fleeces to woolroll. Cheap brooms that wore out very quickly. By the time we cut out, I was almost sorry to leave.

Gill started his car and we wheeled off down the rough track. He was happy. He'd just finished his foremanship. Now he'd be granted a probationary stencil and be off to class a shed on his own. "You're over the first hurdle, though, Wade. First year's always the hardest. Now you can go back to the stores for a bludge till second year, eh?"

A cloud of dust plumed up behind us. Steep ridges cloaked in purple Salvation Jane weed pressed in on both sides. Stones flew up from the speeding wheels and battered the underside of the car. It was Thursday. Lunchtime. "Home by eight," we told each other and settled down for the long drive.

And six hundred and fifty kilometres farther south a postman on a pushbike squeaked to a stop, dismounted, propped his bike against the letterbox, and strode across the lawn towards the front door. He had a telegram to deliver.

2

The Long Run

THE TELEGRAM WAS lying on my bed when I got home. I wasn't fond of telegrams. I associated them with uncomfortable beds. But it had to be opened. *WOOLTANA, WILGENA, NORTH WELL. STARTS MONDAY. ADVISE IF AVAILABLE. SMITH. SOUTH AUSTRALIAN INSTITUTE OF TECHNOLOGY.*

Three of the best school sheds in South Australia. All owned by the same family. Three months in the bush.

Wooltana, near Leigh Creek. Ten shearers, 24,000 sheep. Wilgena, near Tarcoola. North Well, near Kingoonya. Twelve shearers, 28,000 sheep at each shed. Working under Hughie MacIntyre, one of the legends of the Australian wool industry. Too good a chance to miss.

"What about Dalgety's?" Mum was worried.

Next morning I showed the telegram to Ray Hanks, the foreman at Dalgety's. "Go for it," he said.

At 5 A.M. on Sunday a blue Volkswagen pulled up outside: my ride to Wooltana.

Big schoolsheds are completely different from small ones. The pace is quicker. There are schoolies rushing about everywhere. Plus camaraderie, rivalry, arguments, assistance and plenty of hard work.

☆ ☆ ☆

Wooltana sits on the flat at the foot of the Flinders Ranges. Isolated but beautiful, with sparkling clear air and the silence of the ages hanging in the atmosphere.

Twenty-four thousand sheep take some shearing: ten shearers, two woolpressers, three board rouseabouts, a penner-up, and six schoolies in the woolroom. All first or second years except one, Bill Bagley. He was the foreman and one of the real gentlemen of the wool industry.

There were two shearing boards, facing each other but separated by the sheep-holding pens. Six shearers on one and four on the other. Two board rouseabouts patrolled the six-stand board. A lone rouseabout handled the four-stand. Two woolrolling tables, with two schoolies on each. Another schoolie collected, trimmed and graded the wool shorn from the sheep's bellies and the fleece trimmings which poured from the rolling tables like log chips from a wood-chipping mill.

The woolrollers also pulled out the wool that had been shorn from the back of the sheep's necks. This was harsh stuff, dried out by constant exposure to sun and rain and thickly nested with twigs that had lodged in the wool as the animals foraged under bushes for food. The last schoolie collected and graded this neck wool. Whenever he had a spare moment, he helped to clear the six-stand woolrolling table as well.

The six schoolie positions were rotated at the end of every two-hour run. Each schoolie was fired with a determination to wear out as many brooms as possible — and the floor was usually spotless.

The shearers really piled the pressure on. Most of them were guns: top performers, regularly turning out 200 sheep a day. Even with big, heavy wethers they were churning out 160 or 170 each. Long-John Thomson from Broken Hill. Donny MacQuire from the south-east. Dick Ruscott and Chilla Evans from Longreach. They all struggled to keep the ringer's position: top gun. But it wasn't easy. Competition was stiff.

Ten shearers going flat out was a stirring sight. Athletes of the board. Sweat dripping off their bodies. Muscles tensing and relaxing. Shearing arms moving quickly and fluidly. Sheep under complete control, never in any one position for more than a few moments until they were pushed into the let-out chutes, transformed from clumsy, drab woolly bundles into blindingly white, fit-looking, deerlike animals. Still clumsy, though.

Stirring, too, to be part of such a big team. There were no reserves, no passengers. Everyone had a job to do. No room for bludgers.

But team spirit can be a shaky thing. Deep undercurrents of distrust sometimes welled up and washed it away . . .

☆ ☆ ☆

Saturday night. Background sounds of Slim Dusty singing from a well-worn cassette player. Low murmur of voices from a couple of rooms. Raucous laughter from a crowd in the room at the end, shared by Donny MacQuire and Long-John Thomson. Donny was a young shearer, twenty-five or so, while Long-John was in his forties; they were often team-mates. Tonight, though, they'd both had a few too many beers.

After shearing a few hundred sheep and travelling a few hundred kilometres down through old bush reminiscences, someone mentioned strikes and the laughter died down.

Serious business, strikes. Emotions run deep. Then someone else uttered the word that incenses unionists everywhere: "Scabs" — unionists who've worked during a strike.

Through the masonite walls came the sound of raised voices. A scuffle. Heavy bodies pushing and shoving. Bad language. Scraping chairs.

"Righto. Out on the flat. We'll see who's a bloody scab!"

The door abruptly swung open and a patch of yellow light bled on to the verandah. Inside the cigarette-smoke filled room, a crowd of angry faces. Glossed with sweat. Befuddled in alcoholic haze. *Chinkonk!* An unsteady foot kicked over an empty beer bottle as the crowd pressed through the door and out on to the flat to see the fight.

Snap-cold air. Brilliant stars overhead. Great black bulk of the ranges silently blocking out half the western sky.

Donny and Long-John. Drunk. Squaring up. Circling round in the patch of light spilling through the open door.

Shadowy figures out on the fringe of the light. Rouseabouts and schoolies, drawn from their rooms by the commotion, to be greeted by harsh words from one of the shearers: "Piss off. None of your bloody business."

Sudden flurry of punches. *Clonk!* Donny stopped a beauty. *Smack!* Long-John's head snapped back. *Thump!* Donny's head deflected another blow. *Thumpsquelch!* Long-John's nose squashed against his cheekbones. Spurt of blood. Startled retaliation: *Smack! Smack! Thump!* Donny's eye went black immediately.

Then old Tim, the expert, appeared. "Bloody hell, you blokes! What's going on?"

Long-John was angry. "The bastard called me a scab!"

Donny wasn't going to keep quiet either. "Bastard shore during the strike." That was back in 1956!

"For old rates!" Long-John was almost pleading.

Those 1956 strikers were protesting against a drop in the shearing rates. Wool prices had fallen. Woolgrowers wanted to bring the shearing rates down to compensate for their losses. The Arbitration Court thought that was a good idea, but the shearers vowed to fight the decision to the end. They decreed that only woolgrowers who paid the old, higher rates would get their sheep shorn by union shearers.

As usually happens in cases like this, people on both sides suffered a mauling.

The strike ran from January to October. Union levies and donations helped feed shearers' families deprived of their income, but for some that wasn't enough to ensure survival and they shore sheep for the new, lower rates. Woolgrowers, deprived of their own income and seeing the wool on their sheep growing to virtually valueless lengths, were glad to hire them.

Eventually the shearers won. But deep, bitter rifts had been carved through their ranks, and the bitterness had been passed from one generation to the next. Donny had still been at school during the strike. But his dad had once told him that he thought Long-John had scabbed. And a scab was a scab for life.

"Behaving like bloody kids!" old Tim declared. "Go to bed or I'll sack both of yers. And that's no bull!"

Next morning, Donny and Long-John were back on the board, scarred but sober. Mates again.

Life in the woolroom was hard and fast, but it was becoming enjoyable. All the schoolies were good blokes.

Bill Bagley's kindly and skilled leadership moulded us into a team. Everybody pulled their weight. Hughie MacIntyre willingly passed on his knowledge and we all felt that that we were really getting somewhere.

Wooltana's shearing went without a hitch and three weeks from the time we started we were on our way across to Wilgena, where there would be another two shearers, one more presser, an extra rouseabout, as well as two more schoolies. The pace was about to go up a cog.

RATTLE YOUR DAGS

THE DESERT WAS blooming. North-west South Australia was enjoying a bumper season of rain. Red and green carpets of Sturt Peas. Lush saltbush. Belly-high feed for the sheep.

We pulled into Wilgena on a Sunday afternoon, just when the first sheep were being driven into the shed.

"What's that noise?" It sounded like a thousand pairs of castanets all being played at once.

Consternation reigned briefly. Then somebody twigged. "Look at the bloody dags!"

Big clumps of dried sheep manure hanging in the sheep's crutches were rattling as the animals walked about. The manager of Wilgena had gambled and lost. He hadn't had his sheep crutched this year. "Didn't expect so much feed to be around," he told us. But his sheep had made the most of it.

Crutching is normally done between full shearings. A small team of shearers is called in and they shear the wool away from the sheep's crutches. That keeps the wool short so that dags can't build up. But these sheep!

Rumblings started right away. "We'll 'ave to get more money

for shearin' these." That was Donny.

"Too bloody right!" Long-John added.

Next morning the shearers drew lots to decide their position on the board. Chilla drew stand number one. That also made him the union rep for the shed. So he was the one who had to front the manager for more money.

They met out in the yards, overlooking the milling, rattling sheep. It was a justified claim. Chipping away tightly packed sheep manure is not pleasant work. And it's hard on the shearers' tools, the combs and cutters. Chilla didn't have to put much of an argument. He was only being reasonable. The manager was reasonable too — we saw nodding heads.

Chilla walked back into the shed looking satisfied. "An extra comb an 'undred." The other shearers muttered approval. They'd be given a new shearing comb for every hundred sheep they shore.

Shearing started. The mighty diesel motor took up the strain and the shed settled down to the *thump-thump-thump, buzz* and *baaaa* of a big shed.

By the end of the day the woolroom was knee-deep in woolly dags. We had baskets full of them. Woolpacks full of them. They'd consumed most of our smokos and lunch hour. Often the dags came away from the sheep with some good wool on them, and we had to trim it off so that it wouldn't be wasted.

This was a time-consuming, fiddly job. We schoolies weren't getting paid extra, so we complained to Chilla. He was the union rep, and we had to pay fees to be in the union also. But Chilla wasn't very sympathetic.

"There's enough of you rouseabouts to 'ave yer own union rep. It's none of my business!" Young lads don't have any clout in union matters. We didn't get any extra money although we certainly worked a lot of extra hours.

☆ ☆ ☆

But there was still time for fun . . .

Tarcoola, in South Australia's north-west, was the site of a government gold-stamping battery earlier this century. When the gold ran out the battery was shut down and left to rot. Everything too big to move was left where it stood, along with an array of the smaller, more personal trappings of the battery's heyday.

We spent many weekends fossicking through the old ruins. Some of the blokes even managed to find a few specks of gold in one of the old tailings dumps.

Mainly, though, it was the rusting mementoes of years gone by that were most eagerly sought: old tobacco tins, pocket-knives, belt buckles and even an old revolver gradually accumulated in the quarters during the weeks we were in the area.

One Saturday afternoon, while a dozen or more of the team were digging and scratching around the rubbish dumps, one of the shearers made a startling find in a derelict hut. In a wooden packing crate, dusty with age, he came across twenty or so sticks of very old, sweaty gelignite and a couple of metres of fuse-cord.

The sages warned against touching the jelly, which they assured us was unstable, but that didn't prevent a couple of the blokes from taking the fuse outside and lighting a piece of it. Sure enough, despite the decades of decay, it sparked into fizzy life, then defied all attempts to stamp it out.

One of the woolpressers had been a powder monkey on the mines at Broken Hill; he assured us that if the box of jelly went up, it would completely demolish the old shed and leave a very big hole in the ground. After that pronouncement, the more cautious team members took care to place themselves a considerable distance from the place.

Then one of the rouseabouts suggested we should load some of the lethal stuff on to a sheep and make sure it got into the pen of a certain unpopular shearer.

The idea was dismissed, though, on the grounds that when it went up, we'd *all* be covered in blood and guts. But as a result of that suggestion someone's mind was infected with the germ of an idea.

Saturday nights in big contract sheds are often jolly occasions. After a day of rest, with no need to get up early the following morning, the blokes can afford to sit around the messroom late into the night, drinking and yarning, sometimes singing and loudly debating everything from sheep to politics and back to sheep again.

On this particular Saturday night we had a blazing log fire suffusing the messroom with cosy warmth. Grog was flowing freely and hundreds of sheep were being shorn and reshorn. A couple of blokes were playing crib and I was getting down to some overdue letter-writing. About two-thirds of the twenty-four-man team was in the room.

Then the door opened. Only a few centimetres: most of the blokes didn't even notice. Bright flame spurted briefly from just outside the door, then a squarish object, leaving a trail of sparks, was lobbed into the middle of the room and the door slammed shut. One bounce and the object, still spluttering and sparking, slid across the floor and spun to a halt just underneath one of the long trestle seats.

Sixteen pairs of eyes, bulging with horror, stared at it. The source of the sparks was a length of fuse-cord quickly burning its way down towards a dark box-shaped object fitted at one end with two spiral brass contacts.

"It's a bloody bomb!" someone shrieked.

Long-John was first out. He chose the door. Travelling at about the speed you'd expect from an artillery shell, he covered the hundred metres or so across the flat to the water tank oblivious of the fact that his stockinged feet were collecting more and more prickles with each gigantic stride.

Back in the mess, pandemonium was the standing order. Ray, the ex-powder monkey, wrenched open the window, leapt through it, and immediately started a mini-stampede among the others nearer the window than the door.

Muff, a diminutive shearer, was somewhat drunk. With great aplomb he walked into the pantry at the far end of the messroom and blithely closed the door behind him.

I left my letter-writing in mid-word to join the throng which was intent on putting as much of Australia as possible between ourselves and the messroom.

"Get down! Get down!" yelled Dave, the Vietnam veteran.

Nobody needed a second prod. Soon we were all prostrate in the dirt and prickles.

Silence.

No explosion.

Minutes ticked by.

Someone, shivering from the cold, cautiously stood up. Others followed. Ready to hurl themselves to the ground in a trice, a few brave souls crept back a pace at a time towards the mess.

It was Muff who defused the tension. From the messroom came his plaintive cry: "Gord struth, you mob of pansies, it's a bloody battery!"

He appeared at the window holding the "bomb" complete with the stub of charred fuse still smouldering against the brass contacts. Someone had cut the red outer jacket from a six-volt Eveready lantern battery, attached a length of the fuse from the old mine site, and thoroughly convinced all of us that it was a real, live bomb!

Long-John was particularly keen to find the culprit. As he hobbled painfully back from the water tank, stopping at almost every stride to remove deeply embedded prickles, he was contriving all manner of ghastly retribution.

The ensuing witch-hunt didn't take long. We found Pud Dodd absolutely helpless. He'd doubled up with laughter and collapsed on to the ground just outside the messroom door.

Pud was one of the other apprentice woolclassers in the team. He was always playing practical jokes, and this was his best effort so far.

The success of the stunt, however, was lost on most of the victims as they carried him purposefully down to the water trough. Without a glimmer of remorse, they dumped his struggling body into the frigid water.

Honour satisfied, the mob trooped back to the warmth of the mess, leaving Pud, shivering but unrepentant, to squelch back to his room in search of dry clothes.

And so it went. Days rythmically blending into weeks. Shed life becoming ever more routine and uneventful. Most weekends coming and going with little more than the memory of a welcome sleep-in to mark their passing. But not every weekend . . .

TESTING TIME AT THE KCG

"HAAAARRRZZZAAATTT!"

Splintered stumps cartwheeled across the red gravel, almost taking the wicket-keeper with them. Everyone could see that the batsman was out — but it was probably just as well the bowler had let out a shout. One of the umpires was rolling a smoke. The other was taking a swig out of a beer bottle.

The batsman stoutly refused to move. "It was a bloody no-ball!"

The umpire with the bottle wiped his lips and gave the verdict.

"Yer's out, Harry. Piss off." Harry went.

We were at the Kingoonya Cricket Ground. Few towns have their cricket grounds as close to the centre of town as Kingoonya does. In fact, the KCG is the centre of town: it's the wide gravel area which serves as Kingoonya's main street.

On one side there was the pub and general store. On the other a line of weatherboard houses used by railway workers and school staff. In the middle a concrete pitch, covered, on the day of a match, with copra matting.

Spectators sat on their own verandahs, on the front verandah of the pub, in the local phone box, or in their cars. Any of those seating areas offered an unrestricted view of the game.

And cricket in Kingoonya was serious business. Whenever a large shearing team was in the district, the Kingoonya Cricket Club threw down the gauntlet. As most of the sheds in the Kingoonya area are big ones, there was usually no problem in raising a cricket team from among the blokes.

Both teams shared the same dressing-room — the front bar of the Kingoonya Hotel — but rivalry was keen and there was the occasional heated argument in the dressing room after a controversial umpiring decision.

Most of the town, and station hands from within a hundred kilometres or so, turned out to watch the spectacle. A four, a wicket, a near miss, a wide — almost anything, really, was greeted with a raucous chorus of cheers and car horns.

On this day, they'd come to see Kingoonya Town play Wilgena Station shearing team.

The Kingoonya team members were all semi-permanent residents, so they got to play and practise together for most of the year. Many of them were half-respectable cricketers.

And the cricketers in the Wilgena team? Well, they'd been thrashed the year before, but we were sure we would wipe out that humiliation. After all, we had Kenny Dingle-Allen!

We hadn't actually seen Kenny play, but judging by his impressive command of cricketing terminology and tactics, to which was added the seemingly endless list of top batsmen and bowlers he claimed to know by their first names, we were sure we had a match winner.

There was no doubt, Kenny had told us, that he would have played for Australia, but on the day the selectors rang him he'd had a blue with his wife. She'd answered the phone and told Don Bradman, or whoever it was calling, to "piss off". On the strength of that we made Kenny captain.

Time to start. A spinning silver coin flashed in the sun. "Heads!" Kenny called. It was tails. We were sent into bat.

Kenny didn't want to open the batting.

"Prefer to bat down the order a bit . . . let the openers knock the shine off the ball . . . then I'll go in and carve up the bowling. Won't take long to knock the shine off out here — the gravel will chew it to bits."

Kingoonya's opening bowler was pretty quick. Davo, our opening bat, was still staring down the pitch and tapping the crease with the end of his bat when the first ball slammed into the wicket-keeper's gloves. "Jeeeeezus!"

Next ball. *Fzzzzzzzzzzzzzz*. Snick! "Aaaaarrrzzzzzaaaatttt!"

Davo still hadn't raised his bat in anger but the bowler had blasted the ball through and touched the edge of the bat. And the fat bloke with the gloves had taken a clean catch.

The umpires didn't really see what had happened but they were convinced by the enthusiastic appeal. "You've 'ad it, Davo!"

"Thank Christ!" Davo said, and ducked off nimbly in the direction of the bar.

I was next. When the bowler turned to begin his run up, he was almost out of sight. By the time he'd reached the end of his run I'd shuffled two or three paces backwards to give him a clear shot at the stumps. Arm flung back. Wild-eyed stare. One last gigantic stride. Arm whipped forward. *Fzzzzzzzzzzzzzzzz.*

If I could have seen it coming I would have ducked. Only the faint smell of burning leather told me that the red missile had passed within a whisker of my nose.

"Well left!" shouted the wicket-keeper from his position about thirty metres behind the stumps.

"Attack the bastard, Hewsey!" shouted Mick, the surviving opener. When the next ball speared down the pitch, I managed at least to get wood on to it. But it made no difference. The cannonball barged the bat out of the way and scattered the stumps.

Four balls. Two wickets. No runs. I passed Kenny Dingle-Allen about halfway back to the pavilion. He looked like a man with a mission.

Two balls left in the first over.

Ken crouched over the bat and when the demon bowler hurled the ball down the pitch, he was ready for him.

With his feet firmly planted Ken swung the bat. There was a mighty *clonk* and the ball soared skywards. There was no doubt that it would be four. Maybe six.

Just then, a ute pulled out from in front of the general store and accelerated across the outfield. A few of the fieldsmen shouted abuse but the oblivious driver kept going. He didn't even slow down when the ball thudded on to the roof of his cab and bounced into the back of the vehicle. With dust pluming up behind him, he roared off in the general direction of Coober Pedy, apparently unaware of his cargo.

Kenny had run about seven or eight by the time the Kingoonya captain gathered his wits to call: "Lost ball!"

The umpires disagreed. "It's not lost, mate. We all know where it is — it's in the back of that bloody ute!"

Kenny had run about fifteen or twenty by the time the Kingoonya captain realised that the umpires were not going to be swayed. Another vehicle was soon speeding up the track in hot pursuit of the ute. Kenny and Mick kept running.

At thirty-eight their run had dwindled to a trot. Meanwhile, most of the Kingoonya team drifted off towards the shade and the bar.

By sixty-five, Kenny was tottering. At eighty-seven he collapsed. "Get him a runner!" somebody yelled.

That's when the pursuit car bounced into sight. There was probably still time for a quick single, but Kenny didn't bother. The vehicle skidded to a stop in a cloud of dust and the driver lobbed the ball to the wicket-keeper. Kenny's score stood but you couldn't say the same for him. He retired "tired".

It was a brave effort, but it didn't win us the match. Even with the result of Kenny's mighty swipe, we could only manage a hundred and three. We spent the rest of the day retrieving balls from the longest boundaries on the ground. Kenny's reputation remained intact, though. If only Don Bradman could have seen the game.

Back in the pavilion the Kingoonya captain made some gracious remarks about the visiting team and bought us all a beer.

SMITHY AND OTHER VISITORS

LATER THE following week it was our turn to host a visitor to the shed. One we'd all being looking forward to with a good deal of apprehension.

"Watch out! Here comes Smithy!"

Smithy was the shed examiner. He'd arrive unannounced to spend a day or so watching us work and putting us through our paces as prospective woolclassers.

Smithy didn't say much. Not to the schoolies, anyway. He had lots of furtive discussions with the classers. "How's Farmer doing? What about MacPherson? Will Grogan drop out?"

Smithy missed nothing. He'd come hard up against the end of

the rolling table to see how the skirting was being done. If someone tore off too big a piece of fleece wool, he wouldn't say anything. He'd just pick up the piece and offer it to the classer with an impassive expression.

He'd fossick through the locks, pulling out larger pieces of wool that shouldn't have been there. If he couldn't find any larger pieces or lumps of sheep manure, he'd keep fossicking until he did. Then he'd walk halfway across the woolroom to put them in their proper place.

Then there was testing time on the classer's table. Each of the schoolies would be called in turn to spend ten minutes or so classing fleeces under the watchful eyes of the classer and Smithy. Towards the end of the spell the pressure would be poured on. Smithy would hold back a few fleeces until they'd built up into a sizeable backlog. Then he'd start rapid-fire dumping them on the table.

Fleeces were still streaming in from the rolling table as well, so the schoolie was soon wallowing in a sea of wool. It was time for quick decisions.

Look at the length of the fibres; full length? A touch too short? Too long? Check the physical strength of the fibres; are they sound? Is there a weakness? What about the colour; pearly white or a touch too much yellowing? Grease content? Dirt content? How much wool is really there? How many hanks of yarn will half a kilogram of it spin? Sixty-four? Sixty? Fifty-eight? That's judged by the size of the crimps in the fibres. The finer the crimps the more hanks can be spun.

Pass or failure for the year depended to a large extent on most of those decisions being right. If they were, and the classer confirmed that it hadn't been a fluke, the schoolie could look forward to advancing at the end of the year.

When he'd seen enough, Smithy just said "See you!" and disappeared as swiftly as he had arrived. All you could do was hope that there were no black marks against your name in his little red book. But you never knew before the end of the year.

☆ ☆ ☆

Some sheep, of course, are born with black marks.

Black wool is anathema to the wool industry. Shearers dislike it because black sheep have black skin and they can't see where their cutters are going. They also have to call out "Black wool!" whenever they come across a small spot of the black fibres on an otherwise white sheep. Rouseabouts are then supposed to remove the black wool and put it securely in a bin of its own.

Woolclassers hate black wool because black fibres sneaking through in their lines can bring about a severe rebuke from the Wool Commission. If it gets through the woolbuyers' notice, more problems are caused because black wool will not take dye. Finest pastel-coloured flannel spoiled with black fibres is not easily sold.

So black wool is a problem for everybody in a shearing shed. In the past, all sheep with black wool on them were marked at shearing time and ruthlessly culled. Then, in recent years, when someone rediscovered spinning wheels, black wool has fetched a premium price from the arts and crafts people. Today it is quite common to see small flocks of black sheep wandering around homestead paddocks.

Visitors to woolsheds are often bemused by the fuss a call of "Black wool!" can produce. One such person was a portly matron from Toorak who arrived aboard a tourist coach while we were shearing at North Well, a stone's throw from the main Port Augusta to Alice Springs road.

When shearing's on at North Well coach captains win extra brownie points by wheeling their passengers into the shed. It's a real eye-opener for people having their first glimpse of outback life. Few had any idea of the frantic pace of a big shed.

The arrival of a coach was usually announced by one of the pressers shouting: "Hawker!" Nothing to do with travelling salesmen. Just shearing-shed code, warning the blokes to stop swearing because there were women coming into the shed.

The busdriver would herd his mob of people through, past the press, into the woolroom and down the board. If there were any good-looking women among them, the pace of the shed generally went up about two cogs.

On this particular day, the lady from Toorak attracted the

attention of Bob, one of the three wool pressers in the team. Bob was born and bred in Collingwood — and Toorak was definitely not his domain. But he sidled up to the wealthy old dowager and declared himself a fellow Victorian expatriate.

An already over-ample bosom swelled and puffed up as she struggled to assert an air of superiority over Bob. But somehow he managed to look authoritative, almost regal, in his blue singlet, baggy-legged black shorts, Collingwood footy socks and lanolin-saturated desert boots.

He chatted away to her as though they had just met outside a Toorak boutique.

Then someone yelled: "Black wool!" and curiosity got the better of the dowager. "Whey do they call 'black wool' leyk that? What do they *doo* with it?"

Deadpan, Bob explained in his most matter-of-fact tone. "Black wool, madam? Well, it goes to maternity hospitals."

"What on earth for?"

Bob turned to a full woolbin, stooped, and, as he headed for the press with an enormous armful of wool, gave her the answer.

"They use it to make little wigs, madam. You know — until the real stuff grows again."

☆ ☆ ☆

Just before one coach arrived at North Well, Dikko's nose started to bleed. He was one of the schoolies and he was woolrolling at the time. Flat out. The blood was really pumping round his body and pouring from his nose. All over his chest and arms, blood mingled with copious sweat.

There are no reserves in shearing teams, so if someone goes down with a serious illness, it stays that way until a replacement can be sent up. No one stops just for a bleeding nose.

By the time the tourists walked in Dikko looked as though he'd had his throat cut.

"My God!" one of the purple-haired old dears said when she saw him. "Do they always work you like this?"

Dikko didn't miss a beat. Hands a-blur, tearing wool from the fleeces, pulling out the necks and backs, rolling the fleece and tossing it over on to the classer's table. Drops of blood flew everywhere.

"Oh no," he told her. "This is our rest day. Tomorrow we have to go back to working flat-out again."

When we finally cut out at North Well, we'd been at it for thirteen weeks. I'd done eleven weeks before that. Instead of the compulsory four weeks' shed-time first-year schoolies had to do, I'd completed twenty-four.

Some of the blokes at the woolstores hardly recognised me when I walked in.

"Jeez! Look who's here! Thought you must have gone bush permanent!"

Within a couple of days of being back I knew that it was only a matter of time before I would go permanently. After the buzz of a shearing team, the ho-hum of the stores was almost unbearable. But I still had to do my time. Two more years to go.

3

Second Year: Second Prize

SECOND YEAR WAS very much like first year. I even did the Wooltana, Wilgena, North Well run again. And a few other smaller sheds as well. But it was rather flat. The novelty of being a first-year schoolie had gone. The challenge of foremanship and the chance for a probationary woolclassing certificate were still a year away.

Finally that second year came to an end and I was revved up to go out for my foremanship. Schoolie foremen were the classers' right-hand men. They're supposed to be able to run the woolroom, class wool, do the paperwork, help train the new schoolies and generally act as woolclassers without being given the official title.

Where would my foremanship be? Wooltana, Wilgena, North Well was what I wanted — but that plum went to a three-year rival from the Barossa Valley.

We'd been neck-and-neck in first place right through the wool course. I'd topped the theory year. We'd both passed with credits in first and second years. We'd both done our first and second years at Wooltana, Wilgena and North Well and I thought I had him pegged. But I didn't. He was sent back to do his foremanship there.

So in January I was wondering where I might go. Somewhere up north I hoped. Big contract shed. There were bundles of them up there.

I didn't have long to wait. It was a sweat-dripping day in January. All the windows in the woolstores were open but not a breath of breeze fanned Port Adelaide. I was classing in the bulk class when the tannoy speaker squawked out a noise that bore a faint resemblance to my name.

Through the hot, black bakelite phone the bloke from the school asked: "Ready to do your foremanship?"

"Of course. Where?"

"Second prize: Wide Plains outstation, Clay Flat. Forty-eight thousand sheep, eight shearers."

"When?"

"February the ninth."

Wide Plains in South Australia's far north was a big station. Two shearing teams were used. They worked in two sheds, kilometres apart, one near the homestead, the other at the outstation — a sheep station within a sheep station — called Clay Flat.

The sheep were shared roughly equally between the teams simply by being moved to whichever shed was closest.

It sounded like a reasonable place to do a foremanship . . .

☆ ☆ ☆

A stinking hot February day.

Derek, the woolclasser, and me, travelling in Derek's car. Shirts stuck to our backs after the long, weary drive from Adelaide.

Nightfall gave us relief from the direct effect of the baking sun but the air, too exhausted to move around, couldn't work up any sort of a cooling breeze.

We'd left the bitumen hours ago and were now rattling out across the plains on a corrugated two-wheel track.

Dead mulga trees whizzed past, ghost-white in the headlights. Black skeletons stood further back off the road.

Another car had passed this way not long ago. Its dust still hung in the hot air.

Lights glimmered in the far distance.

"Wide Plains homestead. There's the shed. And the quarters."

We still had more than an hour to go.

"Stop here for tea?"

"Why not? It'll be too late to get a feed at Clay Flat."

If it was hot outside, the mess was like a furnace. A female cook had prepared a three-course roast dinner. Droplets of fat seemed to hang in the smoky air. Heatwaves radiated from the red-roaring stove.

A dozen men, drab in dungarees and blue singlets, crammed together at the tables like overweight penitents in a dingy sauna. Three of the blokes were eating with towels swaddled round their heads to stop the sweat dripping into the tucker.

Fly-specked light globes hung down and cast a waxen yellow glow over everything. Hardly anyone spoke. There were just a few mumbled greetings.

Within minutes of sitting down with a bowl of vegetable soup each, Derek and I had gone up to boiling point. Sweat poured down our foreheads, stung our eyes, trickled off our noses, and plopped into the soup.

"Stuff me! This could be a rough few weeks!" Derek was right.

The quarters at Clay Flat were spartan.

A concrete cell block ventilated by glassless windows with no flywire.

The concrete soaked up the broiling heat during the day, then heated the rooms with it during the night. It was like lying in an oven. Sweat-wet sheets. Damp mattresses and pillows. I was dog-tired. Yet I was up before dawn. Into the shower. Tepid, salty bore water.

Out of the shower, I looked across to the shed, tinged with the first rays of the sun, red as radiator bars. The shed was small, with stone walls and a low iron roof. A real sweatbox.

Into shorts and singlet. Say g'day to a few of the blokes as they pass the door. Wander across to the shed before breakfast. Up the ramp, in through the wide main door and into the hot, stuffy gloom. Sheep were stamping nervously. Grinding their teeth.

The paradox was inescapable: this dark, airless, cramped shed at Clay Flat sat on one of the biggest sheep stations in the world. Through it passed half the station's woolclip. Half its income from wool. Enormous amounts of money — just how much depended ultimately on how well the clip was shorn and classed.

You'd think it would be to everybody's advantage to create the best possible working environment for a shearing team. It didn't happen often, though, and I wonder how many tens of thousands of dollars have been shaved from the value of wool clips over the years because men have been forced to work in conditions which sap the resolve from even the most resolute.

And at Clay Flat we had another problem. We were short-handed in the woolroom. On paper it could have been argued that we had plenty of labour: for eight shearers, we had a woolclasser assisted by two schoolies and myself.

In reality it was hopeless. One of the schoolies had never been in a woolshed before. The fleeces were full of sand, which very quickly blunted the shearers' combs and cutters. They spent half their time changing tools. The sandy wool was heavy. When the board rouseabouts tried to pick up the fleeces they usually fell through their arms and ended up in a heap on the floor.

Then they'd be scooped, rolled and shovelled up, and either dumped on the rolling table in a heap or, most often, on the floor at the foot of the table. They were impossible to handle and sort quickly.

Wool rapidly flooded the constricted little woolroom and it was impossible to shift it. Consequently the classer and his staff had to work on the backlog long after the shearers knocked off for smoko, lunch and tea. And all this in the stifling heat. Near enough soon became good enough.

Within a week, tempers were frayed to breaking point. The penner-up was having trouble moving the sheep into the pens. He hadn't equipped himself for the job — he didn't bring a dog. He really needed two or three of them.

On one particularly trying day he snapped, and strangled a sheep. Bulging eyes. Bulging veins in his neck. Bulging biceps. The sheep kicked frantically but the demented penner-up just tightened his grip. When the sheep collapsed, he stamped on its head. Nobody said a word to him.

Then back to the quarters for another sleepless night tossing and turning in a lather of sweat.

The station manager tended to be unsympathetic to the complaints from the shearing team. "You're only here for a few weeks" was a stock answer.

But after a few weeks here, we'd move on for a few weeks at the next place, and then a few weeks at the place after that. Those few weeks added up to about ten months by the end of the shearing year — many of them in trying conditions.

So, Clay Flat was not a happy shed. As the weeks passed, the hot weather continued. There was nothing to do on weekends. Nowhere to go.

Complaints about the quarters increased. It's very difficult for a shearing team to refuse to finish a shed because the quarters are not up to scratch. Rebuilding quarters can take months; major renovations can't be done while the team is living there. But relief was in sight — Easter.

We were a long way from Adelaide but nearly everybody went home for the four-day break. When I arrived at Christies Beach I went straight down to the sea and swam until I barely had the strength to move my arms. Cold seawater sluiced away the tension and the dust. I was tempted not to go back — but I did.

ROODINI AND THE GHOST OF BILLY DARBY

"LOOK! LOOK! THERE'S a bloody great roo!"

Wally Dockerty was beside himself as we drove north-west to the Clay Flat shed again. Fair in the middle of the track was a giant old-man roo. The powerful spotlights on Wally's car had mesmerised the animal and it just stood there, crouching down occasionally as if it was trying to see beneath the blinding beams.

There were five of us in Wally's car, including Derek the classer. "Shoot him! Shoot him!" Wally shouted.

"We haven't got a gun."

"Catch him! Catch him! We'll shoot him when we get to the shed!"

That sounded a reasonable idea, so four of us leapt out and surrounded the unfortunate animal. Derek grabbed the roo's tail — and let out a yell as though he'd been electrocuted; kangaroo tails spend a lot of time dragging on the ground, picking up thousands of tiny prickles. But Derek hung on. Someone else grabbed one of the big back legs. I grabbed the other one. Davo put a headlock on. Wally screeched encouragement. The roo kicked and scratched and snorted but he was overpowered. We had him on the ground.

"Now what do we do with the bloody thing?"

"Tie him up! Tie him up!" Wally ordered.

"We haven't got any rope."

Derek whipped off his tie. The rest of us were not wearing ties, so we surrendered our belts and the roo was trussed up like a Christmas turkey. That didn't stop him kicking, though. Despite his struggles we half-carried, half-dragged him back to the car.

"The boot's full!"

"Put him in the back seat then."

"There're three of us in the back seat already."

"Well, hold him across your laps." More grunting, kicking and snorting.

"He's in."

He might have been in but he certainly wasn't giving in. As we drove on, those mighty back legs pounded the door lining to shreds, but Wally didn't worry.

"Never mind! I'll upholster the whole bloody car with his skin!"

Then came an earsplitting shriek.

"*Aaaaaaaaaaagh!* The bastard's gutted me!"

The roo had sunk his teeth into Davo's navel.

After an hour or so of this sort of action, catching the roo didn't seem to be such a good idea after all. But then the darkened shed appeared in the headlights.

Another hundred metres and we pulled up outside the quarters. It was two in the morning.

"Now what are we going to do with him?"

"Shoot him straight away!"

"Na. If you do that he'll be too stiff to skin in the morning. Let's put him in one of the rooms."

"Whose?"

"Deadrock's!"

Deadrock was the nickname we'd given to a bespectacled first-year schoolie whose real name was Livingstone. This was his first shed.

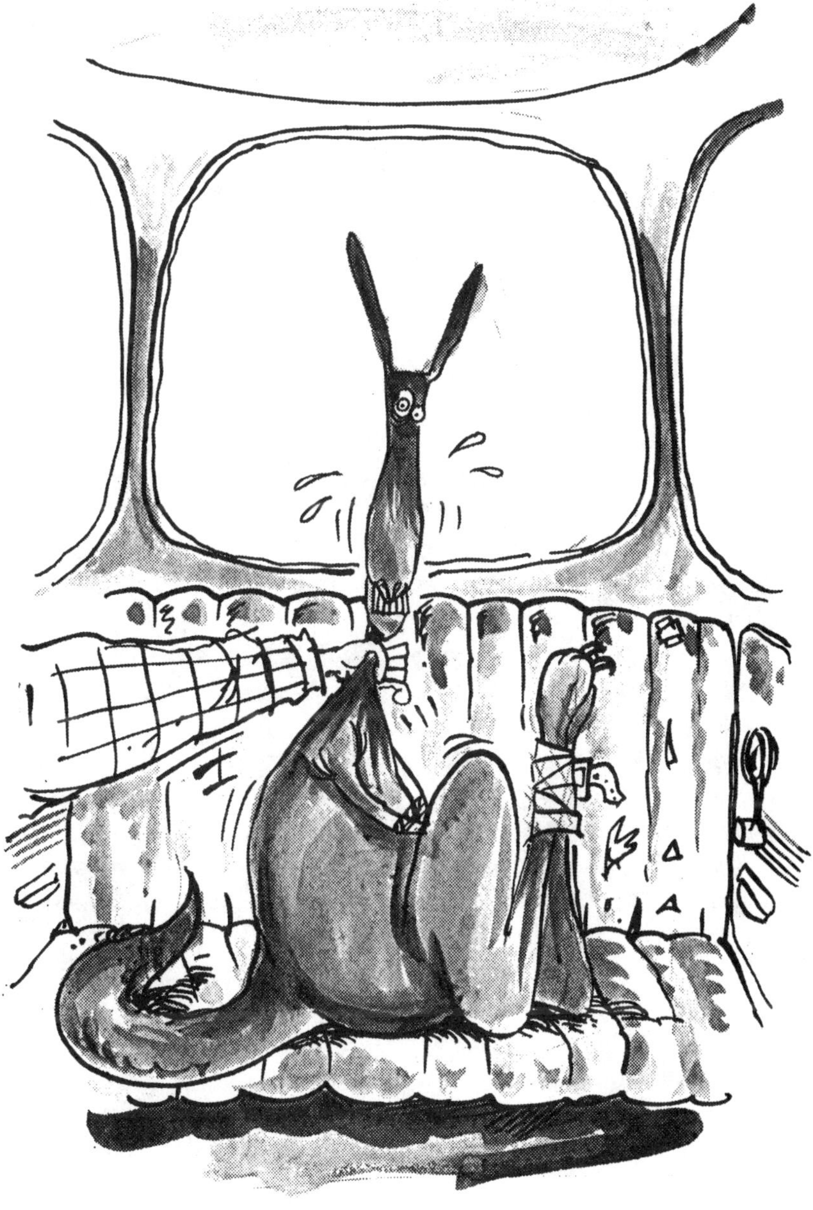

When we found his room, he was fast asleep. He had his glasses off, so he would be short-sighted when he woke. The roo was dragged by the tail down the passageway of the quarters and propped up against Deadrock's door. Wally eased the door open and the bound roo toppled in. A few more shoves and pushes and the furry, furious body slid clear of the door.

"That'll give Deadrock a surprise when he wakes up!"

We didn't realise how soon that would be. While Davo went off to clean up his bleeding navel, the rest of us went off to bed. There was some muffled giggling and Davo made a bit of a racket as he rummaged for a dressing, but the quarters soon slipped into silence.

Deadrock's scream of horror would have done justice to someone being burned at the stake. It was accompanied by fearful crashing and tearing noises. The roo had cast off his bonds and was trying to escape through the open window, which happened to be directly above Deadrock's bed!

The scream woke everybody. Only five of us knew what had caused his distress, but the word soon passed around in furtive whispers.

By the time someone opened Deadrock's door, the roo had gone. The only sign of his passing was the woolclasser's tie lying in a knotted loop on the floor.

"Jeez," Davo muttered, "must have been a real Roodini to get out of those knots."

Deadrock was sitting up in bed, clutching the bedclothes under his chin. He'd jammed his glasses on and his magnified eyes were bulging right out of their sockets. The room was lit by the beams of a dozen torches.

"He bloody well tried to strangle me!" Deadrock was clearly upset.

"Who did?"

"He did! The bloke that was in here!"

"Where did he go?"

"Out the bloody window! Who was it? You bastards, *who was it?*"

"Couldn't have been any of us. We've all been asleep!"

Everybody vouched for each other.

"Must have been the ghost of Billy Darby," Davo said solemnly. None of us had heard of Billy Darby but we all agreed with Davo.

"*Who?*" squeaked Deadrock.

"Billy Darby. He was an old blade shearer who died here one year. He said he was crook but none of the other blokes believed him. Even when he keeled over in the shed they just propped him up in a corner till the end of the run. By the time they'd finished the run, it was too late. There was nothing they could do for him. He swore he'd get revenge, then he just rolled over and died. He's been back a few times now. You must be in his old room."

Davo's delivery had most of us half-convinced. Deadrock believed every word.

"Does anyone want to swap rooms with me . . . please!"

"You must be joking . . .!"

A week or so went by. Most of the team forgot about the ghost story. But not Deadrock. He often turned out for breakfast looking as though he hadn't slept a wink. Then, late one Saturday night, Deadrock laid the tale of Billy Darby to rest for ever.

Syd was drunk. So was nearly everybody else in the mess. The rest of us were in bed. Someone in the mess brought up the Billy Darby yarn again and there was a huge roar of laughter. Whereupon Syd scurried to his room.

Within minutes a woo-hooing white apparition was dipping and weaving across the flat about twenty metres from the quarters, heading for Deadrock's window.

When it got just short of the window, Deadrock heard it. He sat bolt upright, jammed his glasses on, and saw the "ghost".

Fortunately panic gripped him: fortunate, because when he grabbed his rifle and blasted off a shot his aim wasn't very good. The bullet passed through the white sheet, but it was off-centre. Just enough off-centre to miss Syd by a couple of centimetres.

When Deadrock and Syd were finally calmed down, a solemn promise was made in the mess: "No more stuffing around with ghosts, all right?"

Cool nights eventually came. We started sleeping under blankets. Tempers improved.

Smithy came and examined us.

"How's it going?" He didn't have to ask. There was wool all over the floor. But he knew why. As he left he told us all: "Stick to it!"

Then came that magical day. Cut-out. One by one the eight shearers caught their last sheep and shouted: "Sheep-oh!"

But there were no more sheep — and I was effectively a probationary woolclasser. It was an automatic pass from foreman to a probationary stencil.

There was no elation, though. When I walked into my room for the last time and flopped down on my bed, I was too busy thinking how glad I was to be free of the place. And as we drove out next morning, I offered up a fervent prayer that I never be sent to class at Clay Flat.

4

A Shiny New Stencil

WHEN MY STENCIL arrived, I hung it on my wardrobe door. There it stayed for several weeks, shiny and new. I was thrilled. Mum and Pop kept coming in to look at it. It was cut from brown flexible plastic — an outline of Australia with my probationary classer's number in the centre. That was my signature.

The woolpresser would use that stencil to brand every bale of wool that I classed in the sheds. It would give buyers confidence that the wool had been properly prepared. And it would give the Wool Commission someone to hang if it wasn't!

The blokes in the woolstores congratulated me and I pranced around as though I'd been knighted. Then the foreman called me in.

"You want to have a crack at a shed?"

"Er . . . yes . . . where?"

"Meningie, fifteen thousand sheep, merinos. You can put it in for your ticket if you like."

To win a full professional ticket I had to class two shed clips satisfactorily, one merino, the other crossbred. When those clips arrived at the woolstores they would be assessed. A good job meant success.

"Okay. I'll take it. When do I start?"

"Monday week."

I left home on the Sunday evening. Mum and Pop waved me off. So did my sister, who was over from Sydney. Everybody had misty eyes, and I felt as though I was going off to war.

The shiny new stencil was no longer shiny and new. Using fine sandpaper I'd scratched it in a circular pattern that resembled a presser's stencilling action, then rubbed black shoe polish into the scratches. I'd even scratched around the hanging-up hole in the corner. A casual glance at my handiwork projected the impression that the stencil had been hung and used in a hundred sheds. First-time classers can get a lot of stick from the shearing team; I didn't want that to happen to me.

☆ ☆ ☆

The shed sat near a corner of the Princes Highway, just past Meningie. The elderly owners were friendly and showed me to my room in the homestead.

"We've eaten, I'm afraid — but would you like a cocoa?"

"Yes, please."

"Would you like to see some slides of our holidays?"

"Sure."

Cocoa was served. The projector came out. The lights went out.

Three hundred or so of the slides were reasonable. At least they were in focus and correctly exposed, even if they did only show favourite grandkids standing in front of fountains and museums in Brisbane.

The rest were awful. I fell asleep twice, but it made no difference; they just woke me up with an offer of more cocoa and kept feeding slides into the maw of the projector. When the last slide appeared it was two in the morning. I suppose the poor old dears didn't get many visitors.

It seemed that I'd only just closed my eyes when the alarm clock went off. Over breakfast the owner briefed me on what I was likely to expect from his wool, then we went down to the shed.

I strode in as though I'd been doing it all my life. Shook hands with the presser and handed him my stencil. He hung it up without a second glance. Shook hands with the shearers and rouseabouts. Told the rouseabouts I expected them to wear out a broom every week.

"Go easy," said the owner. "Do you know how much brooms cost these days?"

I walked into the sheep pens and inspected a few fleeces on the backs of the sheep. Then I strode into the woolroom and chalked up the wool types over the woolbins. "Should be a breeze."

"Righto, chaps." It was the owner again. "I just want you to know that this is Wade's first shed, so I'd like you all to give him as much help as he needs . . ."

☆ ☆ ☆

Everything went smoothly at Meningie. The owners were better farmers than photographers, and their woolclip proved it. A couple of weeks later I was back on the road again, heading home with a cheque and a bonus in my pocket. I never did work out whether the bonus was for doing a good job in the shed or for sitting through those slides.

So that was the merino clip done, and I had no doubt that it would pass close scrutiny. My main problem now was to find a crossbred clip to achieve my full ticket.

Crossbred clips were as scarce as truthful politicians. Out of all the sheds I'd done as a schoolie, not one of them had been a crossbred flock. Eventually Dalgety's found one for me, in the Mallee down the south-east of South Australia. Russ Mallett's place, near Keith.

Russ Mallett was a top bloke. He'd clawed a profitable little farm out of Mallee wilderness. It was country that had been opened up by a leading finance group and the success of their major development scheme depended on people like Russ and his family. They'd done it the hard way. Started with nothing, mortgaged to the hilt.

The *Women's Weekly* ran a story on Russ and his family once. The picture showed them sitting together in the lounge of their little house on the plain. Two young Aussie battlers with two tiny kids. A giant, untamed land surrounding them.

When I first walked into Russ's shed his jaw dropped. Most cockies worry a little when they know their clip is in the hands of a young, inexperienced classer. He went white when I told him that these were the first crossbred sheep I'd ever seen. Some woolgrowers might have complained to Dalgety's, but Russ gave me the benefit of the doubt. After all, I reassured him, I had classed crossbred wool in the stores. "Great, Wade. Great!"

No room for a foul-up here. No room for a slipshod job.

As usual, the first step was to convince the rouseabouts that if they didn't wear out a broom a week, terrible things would happen to them.

Russ brought a mob of small ewes in first. The shearers flew into them and we soon had a steady stream of crossbred wool pouring into the woolroom. Coarse, greasy wool. Up to thirty centimetres long and wavy. Almost like ringlets sometimes. Quite unfamiliar to me. I experienced a few moments of hidden panic. *"What happens if I stuff it up?"* Then came some rational thought. *"Only two things can happen here. Either you can do it or you can't. And you certainly can't do it if you think you can't."*

I prodded myself into it and it wasn't so bad after all.

A couple of days after shearing started, an old classer from down the road "just dropped in to have a look", then went and had a long chat to Russ. But that was the only display of anxiety I experienced. When we cut out, Russ handed me a $50 bonus.

Back to the slow pace of the stores again. It's cool, quiet and dark between the rows of bales stacked in a woolstore. An ideal spot for a lunchtime snooze. That day, however, sleep was jolted away by the distorted, tinny loud-speaker. I was wanted up on the show floor; the floor where wool is displayed before the buyers bid for it at auction.

My second clip was being prepared for sale and Russ had turned up to see how it looked.

"G'day, Wade." He pumped my hand.

"G'day, Russ." It was my turn to be anxious. The wool valuers were looking over the clip. If they didn't like the way I'd prepared Russ's clip, he wouldn't get a good price and I wouldn't get my professional stencil until I could find another crossbred clip. That could take months.

"Don't worry, Wade. I told 'em what a good job you did."

"Thanks Russ, why don't you tell them again?"

We stood together chatting about his sheep, the weather and the price of superphosphate while the valuers picked and poked their way towards us through the lines of fleece wool.

When they were only a bale away, Russ let out a howl.

"Bloody hell, look at *that*!" He leapt forward to the bale that separated us from the valuers, plunged his hand into the wool and pulled out an enormous handful of mismatched wool.

The bale had been branded AAA CBK: the finest wool in the clip. In Russ's hand was some of the coarsest wool you'd ever see. It was like finding a turnip in a bag of frozen peas.

The valuers swooped.

I nearly swooned. Russ took one look at my face and doubled up with laughter. So did the valuers.

They'd set me up. Russ had planted the coarse wool. Then he'd kept me talking until the valuers reached the loaded bale.

"Jeez, that's worth a dollar a kilo just to see your face," Russ said as he wiped his streaming eyes with his handkerchief.

A week later I got the letter telling me that I'd passed both clips. I was a professional woolclasser. And it was time for a decision: the stores or the bush?

INTO THE REAL WORLD

IN THE STORES? Security. Year-round work. A regular income boosted by occasional higher-paying sheds.

In the bush? More freedom. The chance to travel. The chance to earn more money. But not real security. Away from home a lot. Rough conditions.

I asked the supervisor what chance I had of moving up the ranks to become a wool valuer for the firm.

"Practically none," he told me. "We select our wool valuers from our staff, not our wages blokes." I didn't know if that was true or whether he was politely telling me I'd never make the grade. So I resigned.

Mum and Pop certainly weren't happy.

"You're out of work, lad! We left Britain so you *wouldn't* be out of work!"

But I wasn't unemployed for long. I made a few phone calls to contractors I'd met during my years as a schoolie. One of them, Gary Tribb, had some work.

"Glad to give it to yer!"

"Bloody glad to have it!"

The plan was this. Gary was a big contractor. One of his smaller runs was a string of six or seven sheds that stretched from Port Lincoln on the west coast to just north of Port Augusta. He needed a classer-overseer-expert. Someone who could class, run the shed, hire, fire and buy stores. And be the expert — the shed mechanic.

Experts fix the shearing gear when it breaks down. They change the emery paper on the sharpening discs and sharpen the shearers' combs and cutters.

"But I've never done any of that!"

"Don't worry. There are two weeks before the west coast run starts. Come up ter New South Wales with us as a rousy and I'll teach yer!"

"Errr . . . okay. We'll see what happens."

"You'll be right. See yer Sunday."

Early on Sunday morning we left Adelaide and headed for Broken Hill. At the Hill we turned north, through Stephen's Creek; past the miners' cars parked outside the Creek Hotel while their owners were inside enjoying a Sunday booze-up. On up the road towards Milparinka, we turned off at Williams Tank, then went out towards the shed at Mount Arrowsmith. It was hard country. Early explorer Charles Sturt had learned that.

The shed had been going for a couple of weeks. One of the rouseabouts had left and I was his replacement.

"How am I going to learn all this experting business in two weeks?" I asked myself. Blokes spend years learning to handle shearers tactfully and sharpen their combs and cutters.

Gary was an optimist, though. He gave me intense instruction: how to strip down and overhaul a shearing handpiece. How to set it up again. How to repair and change the drive-belts that transferred power from the diesel motor to the shearing stands. How to get the worn-out emery paper off the discs. How to put new paper on.

Then he tried to show me how to grind the combs and cutters. That's not easy. Well, grinding them is straightforward, once the disc and the stirrup that holds the comb or cutter is set up properly. The difficult part is grinding everybody's tools differently, according to their own special requirement.

"Yer've got to give three heavy pushes, then one light one at the finish," one shearer said.

"One long push is the only way ter do it," another maintained.

"No pushes. Long, steady strokes against the disc. Two of them," was a different request.

"I always get the best cut from three of 'em," a fourth suggested.

However, every shearer agreed that the grinding marks left by the revolving disc had to run straight up and down the teeth of the combs and cutters. That meant finishing the tool off about halfway towards the edge of the disc from the central hub. I spent hours trying to get those marks exactly up and down the teeth. Just a little bit out and there were howls of protest from the wronged shearer. It was a nightmare.

By the middle of the second week, Gary was leaving all the grinding to me. The shearers complained and whinged — I would have, too. Gary just said: "Give the young bloke a go. Everybody has ter learn!"

You wouldn't believe the pressure I felt whenever I went into the engine room to start grinding. There was usually at least one bloodshot eye peering in through a chink in the corrugated iron wall.

"Two pushes on mine."

"Three on mine."

"One long, steady push."

Hang on — that didn't sound right. "You said two steady pushes yesterday!"

"No, I didn't!"

After a couple of days the pressure was getting me down. I was scared of making a mistake, especially while I was being watched.

Then I noticed that, despite my grinding, the shearers' tallies weren't going down. They must have been getting a reasonably sharp edge. I felt it was time to ease the pressure a bit. So I pointed out that the tallies hadn't dropped.

"Ah! That's because we're all working harder to make up for the grinding!" Another approach was needed.

There was a short length of four-by-two timber in the engine room, and next time I started to grind I made sure there was an old comb near the grinding disc. As usual I collected all the tins with the combs and cutters in them. An eyeball appeared in the usual place. I put the old comb on the grinding stirrup, picked up the four-by-two and jammed the end of it on to the stirrup.

When I leaned on the wood, the emery bit deeply into the comb. A thick shower of red sparks fizzed up. The engine lost a few revs and yell of horror went up from the board. The engine-room door burst open and a gaggle of anxious faces peered down on me. I burst out laughing. The shearers didn't think it was particularly funny — but after that they left me alone for a while.

Another week, and it was time to quit the interior for the coast.

THE LITTLE BLACK-AND-WHITE BLOKE

BEFORE I LEFT Adelaide Gary made me sign specimen bank signature forms. Then he gave me the cheque book. And the stores. Just as I climbed into my panel van to drive off he added: "Oh. One other thing. You might need to do some penning-up. Better get yourself a dog!"

"But I don't know anything about dogs!"

"No sweat. Drop in to see George Hill at Bluegum Stud, just outside Lincoln. He'll fix you up. He breeds champions."

☆ ☆ ☆

"Bluegum Sheepdog Stud" was what had originally been painted on the tin sign hanging over the gate. Time and weather had removed most of the lettering, but the crude cut-out of a supposedly alert kelpie stuck on top of the milk-urn mailbox proved this was the place I was looking for.

"You can pay more money for someone else's dog but they

won't be any better than mine," George Hill told me as he hauled a reluctant black-and-white pup out of a tiny pen. He was collie-kelpie cross, and both his parents had been Australian champions, although George "couldn't lay his hands on their papers at the moment". Fifty dollars and he was mine. Although I didn't know it then, I had just bought a dog who, over the next few years, would grow on me like a wart.

With her nose pressed imploringly through the welded mesh of her enclosure, the pup's mother watched as one of her sons disappeared for ever into my van. The pup didn't seem to care much and happily settled down to sleep on the seat with his head on my lap. I looked down at his little black face, peaceful in sleep. A faint blaze of white ran down between his eyes. He was as tiny as a kitten.

"It's going to be a long time before you do any penning-up, pup," I thought.

We, the new dog and I, had travelled just a short distance when he woke, stood up, fell clumsily off the seat and deposited a large puddle on the floor. Small and furry though he was, he got a shaking by the scruff of the neck, his face rubbed in his own piddle and an unceremonious dumping on to the grass at the side of the road.

He showed that he could learn quickly. Next time nature called he didn't alert me by falling on to the floor. An odious smell announced his next effort, this time deposited quietly on the seat alongside me as I drove.

I called him Mowie, after a popular type of fish. He spent the first three months of his life tied to the roo bar on the van during the working day, and sleeping on, or if he could manage it while I was asleep, *in* my bed.

One of the wise old blokes in the team did his best to help train the pup. "Don't make a pet of yer dog, woolclasser, or he'll never be no good!"

☆ ☆ ☆

Port Lincoln is a windswept town with lots of character. Granite outcrops sweep down to the sea. Green paddocks roll back on all sides of the town. There are handsome colonial buildings and plenty of fresh sea air.

While we were working close to Lincoln, I went there every Saturday morning to buy stores. I used to leave Mowie behind but Mike, the woolpresser, often came with me. He was a woolclasser too, but he hadn't been able to get a run classing. So, as he was a big, strong bloke, he'd signed up as presser. We became firm friends, and after shopping we usually had a few beers in one of the local hotels.

On this particular Saturday morning I got the surprise of my life when I spotted one of my old schoolmates from Adelaide, Greg Wilkins. He was a tuna fisherman in Lincoln.

By mid-afternoon the three of us were drunkish. By the time the band started to "test one-two" its way through its tune-up session, we were properly drunk. And still swapping tales about shearing tuna and berleying up sheep.

"Goinferpiss." Greg tottered off towards the gents.

Over in one corner of the lounge was a group of girls. Alone.

"We'll go nget freeof em!" Mike decided. Then he had second thoughts.

"Mmmbettermakeit four, ncase swan sadud!"

So we zig-zagged off to make the conquest.

"Piss off! You're drunk!"

Back to the bar.

"Mussabin somthin you said, Hewsey!"

"I din say anythin!"

"Shoulda spokeup then!" Mike was doing long, slow blinks as he swayed on his feet.

When Greg came back, Mike pointed out the same group of girls. "Greg. See that one in the corner? She want stadance with you."

"Bullshit!"

"Fdinkum! She came overn tol dus. Din she Hewshey?"

"Sright!"

Mike's long, slow blink must have convinced Greg because he lurched off to the conquest.

We could lip-read the response he got. "Piss off! You're drunk!"

Mike and I roared with laughter. Greg did too and gave me a good-humoured push. The same sort of push he gave a 100-kilogram tuna.

I toppled backwards and bumped into the biggest bloke in the bar. Big, scowling face. Thick, black beard. And cross. I'd spilled his beer!

He propelled me back towards Greg.

Like two snooker balls Greg and I collided and Greg cannoned into Mike, who fell over. As he went he tried to save himself by grabbing at the bar.

He missed and grabbed the pocket of the jacket that was being worn by the bloke next to him. That didn't save him. Ninety-five kilograms kept falling and took the pocket with it.

Normally, having a good jacket ruined by a falling drunk would make anyone furious. This bloke probably wasn't pleased, but when he saw the size of Mike he decided against violent retaliation and let him off with a stern, "Crikey mate! Watch what you're doing, will yer!"

Meanwhile, further down the bar, I wasn't doing very well at all. The bloke with the black beard had me by the throat. I was frantically swinging my fists, but his arms were longer than mine and I was just causing a breeze — and running out of breath!

Then the bouncers moved in, just in time I think. They didn't seem interested in my argument that I had only been fighting for air. We were all thrown out.

On the footpath Mike stepped in to rescue me from further throttling, then offered the advice, "you'll have t'learn t'use yer front feet, Hewsey!"

Greg said he was going back to his tuna boat.

Mike and I suddenly felt hungry.

"Fish 'n chips?"

"Good idea."

We got into the van and headed around to the fish and chip shop.

"Plenty saltn vinga, please."

Then we sat in the van to eat them. Well, to start eating them, anyway.

Consciousness slipped away into a deep sleep . . .

Sounds came through first. Cars driving past. On the way to church, probably. Then the pain: raging hangover, stiff neck, sore back. And the smell. The pillow smelled awful.

It hurt just to open the eyes; misted up windows, grey light, and that smell. Then I realised. There was no pillow. I'd gone to sleep with my head in Mike's uneaten fish and chips.

A huge pair of feet dangled over the back of the front seat. Toes down. Mike had crawled into the back of the van and gone to sleep on top of the stores.

When he woke up he obviously felt no better than I did.

"Strewth. You must have to be horribly ill to die!"

Very quietly we drove back to the shed.

Life took on a more sober character after that.

LOW DOGS

FROM OUR FIRST shed near Port Lincoln we gradually worked our way north, up the Eyre Peninsula. It seemed that we encountered something new in each shed.

One cocky had three dogs that would only bark when they were lying flat on their bellies!

When we first drove into the property they kicked up a hell of a fuss, barking and slavering at the end of their chains. Oddly, though, they were all pressed as close to the ground as they could get. Even their heads were half-buried in the short grass in front of their kennels.

Mowie stood up on the front seat and growled at them as we cruised past.

The mystery was solved on our first night. Around midnight, with a bright silver moon starkly etched against the cold, black sky, something disturbed the dogs and they set up a furious chorus. The homestead was only a hundred metres or so from the shearers' quarters and the owner's voice was plainly audible.

"Shut up, you bastards!"

More barking, then — BANG! BANG! BANG! BANG!

BANG! BANG! BANG!

Gunshots whiplashed through the cold night air. Silence.

Then the sound of the owner's bedroom window being slammed shut.

"'E's shot his bloody dogs!" someone said.

But he hadn't. All three were alive and well in the morning. And a close look at their kennels showed that the roofs were riddled with bullet holes.

"Jeez, I'd learn to bark on me guts, too!" one of the shearers observed.

BLOOD ON THE BOARDS

THREE MONTHS LATER we wheeled into Mulo, north of Port Augusta.

Mulo was our last shed in the run. Almost a holiday shed. Four shearers. Clean. Not too quick. Good company. Decent sheep and weather.

We expected to be there for about three weeks but after just one week, disaster struck.

Four o'clock one afternoon, halfway through the last run. Time to change combs and cutters. Time for a swig of water and a smoke. Each shearer finished his sheep, pulled his shearing piece out of gear and stood up. The buzzing stopped. All was quiet, except for the deep pounding diesel and the swish of brooms being worn out.

Deft rolls of the wrist disconnected the handpieces from the down tube. Combs and cutters came off and were tossed into water tins to soak off the wool-grease. New tools were screwed on tight, with a liberal lacing of oil. Handpieces were reconnected to their down tubes and dropped on the board, ready to begin — after a smoke. A few minutes' breather before the last hour, the downhill run.

One fleece lay on the table, one on the floor. The woolroller was finishing off a fleece while I fussed about in the next bin to be pressed. Top line. Just one more check before it went into the press.

Mike the woolpresser was weighing and branding the last bale, silhouetted against the wide-open doorway. There was a smell of new jute from the woolpacks. Outside, in the distance beyond Mike, blue hills reared up out of the red plain. A flock of pink and grey galahs flapped across the doorway, framed against the sky like birds on a movie screen.

One by one the shearers get back to their work. One by one the handpieces start buzzing. Slam! Slam! A struggling sheep thrashed its legs down on to the board.

"Steady up!" Barry Paddick, a quiet shearer with a cool temper.
Slam! Slam . . .! Slam! Slam!
"Hey! Come on, yer bugger!"
Slam! Slam . . . slam!
"*Aaarrrggh!*" A piercing scream.
"Hey! Look out!"

A dropped handpiece, like a chain saw out of control. Heavy metal and gnashing teeth, doing 600 revs. Twisting and bouncing on the end of the down-tube. Waving round like a loose high-pressure hose.

"Barry! You all right?" shouted Trevor, the bloke shearing next to him.

"Bloody hell! Quick, someone!"

By the time I got to the board, Trevor had pulled Barry's handpiece out of gear. The other shearers had stopped. A couple of half-shorn sheep trotted into the woolroom then skidded round in panic as Mike rushed around the corner from the press. There was a big pool of blood on the board. Barry was lying in it, doubled up with pain.

"What happened?"

Barry's struggling sheep had caught him off-balance. Its flailing back legs had caught in the elbow of the down-tube and swung it awkwardly away. When it came back, Barry had lost his grip and the handpiece had jagged up into his armpit.

Razorsharp shears, digging in, slicing deeper. It was a very messy cut and he was bleeding badly. We didn't really know what to do with Barry. Lift him into a car and race across to the homestead? Or leave him where he was and go for help? The second choice seemed better. As soon as the station-owner was told, he could call up the Flying Doctor from Port Augusta.

Mike sprinted from the shed and across to his car, parked alongside the quarters. Dust swirled as he spun his wheels in the gravel.

Meanwhile, a press of anxious faces surrounded Barry. We tried to make him comfortable on the board. Nobody had done first-aid. Nobody knew how to staunch the flow of blood. Barry kept his arm jammed down tight against his ribs but plenty of blood still oozed out from beneath it. It was clotting into a thick lump on the board. He didn't make a sound, but his face was white and creased with pain.

Outside the shed, more skidding tyres clawing against the gravel. Two car doors opened and slammed. Racing feet pounded the steps. Mike reappeared with Ken, the owner. And a first-aid kit.

Ken was breathless. "Meryl's on the radio now calling the doctor."

When he looked at Barry he made his mind up straight away.

"We'll have to stitch it up before the doc gets here or he'll bleed to death. Lift him on to the table."

Four of us picked Barry up and shuffled across the board into the woolroom. Big dollops of blood splattered down on to the floor.

Badly cut sheep get stitched up on the board. Either by the shearer who cut them or a willing helper. It's a simple procedure. Take a thick needle and a length of strong thread. First try to stitch up the big blood vessels. Near enough is good enough. Then pull the cut back together and stitch that up. Finally, pull the skin together and sew that. Make sure this part is neatly done . . . in case the owner happens to see it!

Now the owner was about to do a similar job on Barry . . . without anaesthetic.

Mike lifted the injured arm up over Barry's head. The wound looked like a rabbit with its throat cut. Barry winced at every stitch. Ken's probing fingers were covered in blood. So were his boots. It was a pretty rough job — but it seemed to work. The flow of blood slowed, then shrank to a dribble. Barry had tears in his eyes as he tried to lower his arm.

Ken wiped his hands on his pants and wiped his brow with his forearm. The gesture left a streak of sweaty blood over one eyebrow. "We'll take him over home to wait for the doctor."

It was a slow, painful journey for Barry. First we lifted him down off the table. That brought a stifled *"Aaaaggh"*. Then we had to get him out of the shed, down the steps and into the car.

Light was failing, so Ken sent one of the station hands out to light the flares on the airstrip.

About forty-five minutes later a distant throbbing in the sky grew louder. Two beams of brilliant light speared earthwards. The Flying Doctor circled once, then raced in for a bumpy touchdown on the gravel strip.

We eased Barry upright again and helped him into the car. Ken drove right up to the plane. With a cool, almost detached efficiency, the doc and his nurse gave Barry the once-over, strapped him on to a stretcher, fitted him with some kind of drip and watched dispassionately as he was lifted aboard.

The plane door slammed. Motors coughed then roared into life. Momentarily the pilot's face flashed palely at the cockpit window as he spun the plane around. Then they were airborne. More quickly than I'd expected, the noise shrank away, leaving only the muffled thuds of the diesel lighting plant to puncture the night air.

I turned to Ken. "I'll call the contractor tonight to see if we can get another shearer within a couple of days or so."

☆ ☆ ☆

By this time Mowie had grown into a strapping teenager. Still clumsy, but keen. Every time he saw sheep his ears pricked, his eyes sparkled and he whimpered, begging to be let go. He was obedient, too. He'd sit. Stay. Lie down. Speak up! All on command and without fuss.

Good dogs must have those commands bred into them because I had no idea how he came to learn and obey them so readily. He still slept on my bed and I was firmly attached to the little bloke.

He'd cost me some money, though. The vet in Lincoln had given him a couple of injections. Then Mowie had pounced on a bee. When it stung him he yelped as though he'd been shot and his whole leg swelled up in seconds. More vet fees. After that he chased another bee — for revenge, I suppose — but carefully avoided stepping on it. This one was dispatched with a clomping snap of the jaws. Another piteous yelp. The whole side of his mouth swelled up. Back to the vet.

Then there was the trouble over the shearer's underpants.

Saturday afternoon at Mulo. I was lying on my bed reading. Someone was listening to the races. A few of the blokes were doing their washing.

Suddenly an angry voice roared in the laundry:

"That bloody dog! Come here, you black bastard!"

Mowie might have been obedient, but he wasn't stupid. When someone roared: "Come here, you black bastard!" he left. Quickly.

He skidded round my doorway, tail jammed between his legs,

belly very close to the ground, and jumped straight up on to the bed. Just in time. Merv was right behind him.

"What's up?"

"*This!* And this! And this!" Merv had three pairs of navy blue Y-fronts in his hands. They'd been lying on the laundry floor waiting to be washed. Merv had been out at the line hanging out his dungarees and when he'd gone back to the laundry he'd caught Mowie chewing the crutch out of the third pair. I had to supply replacements.

☆ ☆ ☆

When Mulo cut out, I went back to Adelaide, where Mowie also got into trouble at home. Mum had a poodle now. Chota. Tiny, fluffy and adorable. Bounding with energy. And white.

Now, Mowie thought in very simple terms. Sheep are white. This must be a sheep. Sheep shouldn't be running around by themselves. Especially in the boss's house! He charged. Chota fled but Mowie was quicker.

A mortal yelp — that was Chota. An earpiercing shriek — that was Mum.

When I caught Mowie he had a firm grip on the poodle's left ear. I almost had to twist his tail off to make him let go. Then I cuffed him and he looked really hurt. His eyes said it all . . . *but you're always telling me to "Get hold of em!"*

Mowie spent the rest of the visit tied up to the clothes hoist.

Nor was he the only one in the doghouse. Pop was showing concern at the state of my finances again.

"How long will you be out of work this time, lad?"

"Not long, I hope."

"Will you be taking your dog when you go?"

"Of course!"

Then Bill McGillicuddy rang. He was an old-time classer and I'd spent some time working under him as a schoolie.

"I hear you've got some spare time up your sleeve," Bill said.

"Yes, worse luck."

"Now that you're a real woolclasser, would it be beneath your dignity to come out and work with me as a woolroller for a couple of weeks?"

"Hell, no. I'd be glad of the work. Whereabouts?"

"Er . . . look, it's a bit of a tricky one. The bloke's got lice."

"Great! But I'll do it. Er . . . do you mind if I bring my dog?"

"Not at all."

Sheep lice make wool men very twitchy. They used to be a major pest, knocking thousands of dollars off the value of infected woolclips. Today most sheep are dipped after shearing and this has helped to control the problem. But occasionally some woolgrowers take a punt and miss dipping for a few years, and this can lead to outbreaks of lice.

This particular woolgrower had copped a major infestation and he was going to suffer a big dent in the value of his woolclip.

You could smell the lousy wool when it came through. Harsh. Yellow. Musty. It made your scalp itch just to look at it.

They reckon that sheep lice won't colonise human hosts. But that didn't stop everyone from scratching their hair every few minutes. Mowie slept outside for the first time in his shed career.

I didn't like the way he kept scratching after a day sitting in the woolroom.

We washed the shed down every other night with hot water, disinfectant and insecticide.

When we'd finished the job there wasn't a cleaner shed in South Australia.

☆ ☆ ☆

Back at home again, Mowie scratched constantly for a week. Not lice. Fleas.

He got doused with Malawash a couple of times. This he didn't like at all. After the horror and indignity of the first bath, he refused to come when I called him if I was standing anywhere near the old tin tub I'd used.

When his scratching made it obvious that he'd have to be bathed again, I tied him up first, then filled the bath with the hose. Mowie chewed through the rope and ran off. When I finally caught him he clamped his jaws on to my arm, and yelped when I cuffed him for that oversight — but it didn't lessen his resistance.

Eventually I got the rope tied to his collar again and dragged him, stiff-legged and stiff-necked, back into the garden. He was dog-paddling when I picked him up and dumped him into the tub. As soon as I let go my vice-like grip on the scruff of his neck, he jumped out of the tin tub, shook himself all over me, then raced off to roll in Pop's compost heap.

Chota, who usually smelled like a Parisian fashion model's furry handbag, looked on with disdain . . . from a safe distance.

Gradually Mo stopped scratching but, after a week cooped up in the back garden at home, he was getting frisky, so I took him off down the beach for a good long run.

That's when he discovered seagulls. They were white. They must be sheep. They shouldn't be running around loose. And, after fighting tooth and nail to avoid a bath, he ran full-pelt into the shallows to round them up. Leaping and splashing. Bouncing and barking. He seemed quite surprised when he learned they could fly. He was even more surprised when I heaved a big clod of rank-smelling seaweed at him!

Then another dog-owner walked on to the beach. She had long, blond hair, big, brown eyes and a young spaniel bitch on a lead. Mowie forgot the seagulls and went to sniff both ends of the spaniel.

Ears pricked. Tail up. Proppy, stiff-legged trot. Then, the formalities over, he tried to mount the girl's doe-eyed dog. As responsible parents we prevented the youngsters doing something they might regret later, and introduced ourselves. "Hi, I'm Liz. This is Tippy."

It turned out that Mowie was useful for something other than moving sheep after all!

Dinner. Dances. Picnics. Even met Liz's folks. Unfortunately, they weren't too keen about Mowie. They had a fine pedigree spaniel in mind for Tippy. I didn't totally agree with them: Mowie looked quite noble from some angles.

Liz and I were just warming to each other when a classified ad in the *Advertiser* poured a bucket of cold water over us.

Wanted. Woolclasser/shedworker. Mixed work. Six months. Agricultural/pastoral sheds. Start Monday. Apply Tom Morgan. A phone number followed.

The bank account was nearing rock-bottom, so I had to apply for it. Tom Morgan was a small contractor. Some of his clients had their owner stencils and could class their own wool. That's when he needed a shedworker. Others didn't. That's when he needed a classer. It was less trouble for him to hire one bloke who could fill both jobs as the need arose. I got the position.

The weekend whizzed past. Liz said goodbye with misty eyes and promised to write every day. I had a lump in my throat. Mowie made one last attempt to mount Tippy.

Mount Lofty was just visible against a lightening sky as I drove down Tapley's Hill, through the silent suburbs and city and out on to the Main North Road. Mowie slept with his head on my lap.

5
Things That Count

CLASSING CAME FIRST, with some overseeing thrown in, at a little shed just off the main road, north of Port Pirie.

One day at three o'clock the last sheep for the run scrabbled to its feet and clattered out the porthole. I cut the motor and stepped into the late afternoon sunlight to count out the tally for the run. A forlorn crow *aaaarked* on the other side of the yards. It flew off when I stepped into sight, but soon settled near the sheep-killing block, a short distance away.

Each let-out pen was full of blindingly white shorn sheep — all waiting to be counted.

Counting sheep is a mixture of art, science and good luck. One way is to count all the legs, then divide by four but I could never get that method to work. I found the safest way was to open the gate a little way and try to get the sheep to run through in more or less even Indian file.

Accuracy is essential. There have been some monumental blues over inaccurate counts. Care was needed. So was self-confidence. All the old hands generally count to the very best of their ability, promptly and confidently scratch the numbers down in the tally book, and refuse to budge from those numbers for anyone.

Sometimes, though, there was room for doubt. A tight knot of sheep would bolt for the gate. There might have been seven. There might have been eight. At times like that, if I knew the shearer well enough, I'd ask him, and whatever he said went in the book. On the other hand, if it was a new bloke, or one I didn't get on with, I'd take an educated guess, put the numbers in the book and prepare to weather a storm.

There were, after all, two chances out of three of avoiding a dispute. If the count was right, or over, nothing would be said. Problems arose only if it was under the total.

But there was nothing difficult about this particular count and I was soon on my way to tea and scones back in the shed.

Then one of the shearers, Fred Billings, reared his ugly head. "'Ere, woolclasser! I got forty-four!" he bellowed from the engine room.

He was scrutinizing the tally book with his towel draped over his head like a prize-fighter. When I approached, he glared at me from beneath the cowl and repeated his charge.

There was always a chance that a sheep could have struggled through the fence from one let-out pen to another. "Anybody get one extra?" I asked.

Silence.

"Sorry, Fred, the other sheep must've eaten it."

"I'll bloody eat you! I got forty-bloody-four!"

I was sure he had forty-three and the count stood. But Fred was angry. During the last run he vented his anger by belting a few sheep under the chin with his handpiece. I ignored the first few but he did it again, so I had to tell him to take it easy. No sheep-owner will tolerate his sheep being knocked around. Contractors have lost sheds because shearers in their teams have been too willing to hit the sheep.

Fred didn't like being told off. "If you come near my stand again, I'll give you one under the chin too — make no mistake!"

Tea-time was unpleasant. When two blokes aren't speaking, it makes the cramped mess-room uncomfortable for everyone.

Next morning war clouds gathered. Fred was on his second sheep. *Clonk!* He whacked it solidly under the chin. When I looked down the board, it was into Fred's blazing eyes. The next sheep got the same treatment. Metal collided with bone. *Thud!* And Fred was glaring down the board at me again. So were most of the other shearers.

I took a couple of paces down the board, but Fred was ready for me.

"I told you, you'll get one too if you come anywhere near my stand."

"Grow up and stop belting 'em, Fred."

"Stuff you!"

Five minutes later: *Clonk!* This time Fred had broken the sheep's jaw. High Noon.

All the other shearers stopped, mid-sheep, to see what was going to happen. Fred kept shearing.

On jelly legs I walked down the board and pulled Fred out of gear. He let his sheep go and it struggled to its feet, its broken jaw sagging painfully. Slowly Fred straightened up to face me. That's when I hit him. As hard as I could. Right between the eyes. He went down like a cold beer in a drought. I was terrified in case he got up; I'm no fighter. But he didn't.

"I'll make your cheque up and you can leave straight away," I told him.

Wobbly legs carried me out of the shed and over to the quarters. Twenty minutes later, adrenalin was still racing through my veins as I scratched out his cheque and handed it to him.

"Get stuffed," was the last thing he said as I walked out of the mess.

When we came back for lunch Fred had gone.

☆ ☆ ☆

During this six months Mowie proved himself a valuable ally in the sheds, demonstrating a ferocious willingness to hurl himself into, and against, the biggest, most aggressive wethers.

They can be almost as bad as rams. They'll stand their ground, stamp their feet and even charge a dog with their thick, blunt heads. They didn't bluff Mowie, though. He'd bounce away, then leap forward, teeth snapping. Rough but effective.

Stubborn sheep can drive a man mad. They will pack themselves into a tight corner with their heads down and resist the most feverish attempts to shift them. Those which are dragged bodily from the pack act as though they're attached to a strong rubber band. As soon as they're released, they zing right back where they came from.

On a hot day, with shearers calling for more sheep and abusing the penner-up if they don't get them, it's not unusual to see the odd sheep beaten unmercifully.

Penning-up can become one of the classer's responsibilities, and it is important that there should be no hold-ups. Delays in the sheep pens mean that wool builds up in the woolroom. That piles on the pressure and the classer is likely to make costly mistakes.

Mowie's solution to sheep packing down in a corner was the classic one. He'd race across the backs of the sheep until he reached the corner, then burrow his way to the floor and nip the lowest nose he could find. That did the trick and broke the pack up.

Some of the blokes complained about the bleeding noses. When shearers start shearing a sheep, the animal is held in a sitting position on its rump. Its nose often gets jammed under the shearer's armpit. So the blokes had a reasonable complaint — but Mowie didn't seem to care.

Neither did he reserve his teeth exclusively for the sheep. On one particularly trying day, I vented some frustration on his black, shaggy rump with a well-aimed boot.

Moments later his teeth were firmly implanted in my army-issue sock and he seemed intent on ripping my leg off. When I swung the other boot in retaliation, he let go of the sock and clamped on to my calf. The little black-and-white bloke had spirit. But not everyone loved Mowie.

One woolgrower in particular didn't like his methods; after seeing several of his sheep with nipped noses, he recommended that my dog be either muzzled or shot, or "preferably bloody well both".

He had a point, so I bought a muzzle in the local Elders store, but Mowie hated it. Instead of working, he ran around shaking his head as though he had a flea in his ear. I relented and removed the offensive thing, but only after sternly warning him what would happen if the owner caught him biting sheep again. He seemed grudgingly to concede.

But Mowie wasn't a patch on Black Havoc! That dog belonged

to Pete, a professional penner-up who joined us at one shed near Port Augusta. With Black Havoc as an aid, Pete could easily have made a living rounding up lions and tigers. The dog was the ill-begotten progeny of an Irish wolfhound and something else, was built like a Brahman bull, and had the temperament of a piranha.

Mowie was bluffed. He took an instant dislike to the newcomer. Fangs were bared. Hackles stiffened. Deep dinkum growls. If it had come to a fight, Mowie would have been shredded, so we gave Black Havoc best and I retired Mowie from shedwork for a while. Black Havoc ruled.

He spent most of his time outside the shed. Whenever a shearer yelled "Sheep-oh!" the dog would bound into the shed, bowling anything smaller than a woolpress out of his way. He never broke stride as he hit the pen gates. Long before his owner appeared, he'd hurdled the fence into the sheep race. The number of sheep which needed stitches depended on how long it took Pete to get into the yards.

Pete was always armed with a thick stick, but he never touched a sheep with it. Lumps of fur and skin would fly from Black Havoc's pelt as Pete persuaded him to let go of one sheep and tackle another.

Absolute blind, unreasoning panic swept through the sheep and they flew about the shed like a flock of swallows, seeking refuge anywhere from the coal-black eyes and ivory fangs of this brute.

Pete had only to open a pen gate, and however many sheep happened to be bolting in that direction quickly crammed the pen to overflowing.

It was then simply a matter of prising Black Havoc's jaws from the neck of the hindmost animal and closing the gate.

The blokes reckoned they'd never seen anything like it, but it was all over very quickly.

At lunchtime on the first day the owner walked into the shed, took one look at his torn and terrified flock, another at Black Havoc's Rasputin-like eyes, and Pete was out of a job.

MALE AND FEMALE

TOM MORGAN'S RUN took us out along the Barrier Highway, beyond Olary, towards the South Australia-New South Wales border.

An isolated, distant shed. A shed serviced by an old Bedford mail truck. "Mail truck" wasn't really an accurate name for it, though. "Grog truck" would have been better.

It came out from Broken Hill every fortnight, loaded down with cartons of beer, a few stores and the occasional letter. Old Merv had been driving it since it made its first shiny-new trip. Merv had been in good nick then, too. Now they were both battered old wrecks that leaked a lot.

Merv had spent his entire life in the Broken Hill area. He'd done a bit of shearing, some pressing, some mining, and some time in the local lockup. He was a warm and friendly old bloke but he used to drink himself senseless. He would have given anyone the shirt off his back — but they'd probably refuse to take it because he'd worn it three weeks on the trot.

As mailman, and sort of mobile barman, he was a welcome sight every fortnight. Today was Saturday and mailday.

Steam hissed out from under the radiator cap when Merv shut the motor off outside the mess-room door. Then you could hear a constant drip-drip-drip of oil off the bottom of the motor and the truck door creaking open.

As Merv eased himself down from the cab, he dislodged three empty beer cans. They fell to the ground and, rolled along by the slight breeze, bonked and clonked their way across the stony ground.

"G'day fellers," he rasped, then hitched up one leg of his old, slack footy shorts and drained the contents of his bladder.

When he bobbed up and down to readjust himself, he broke wind: a noise like a toad exploding slowly.

He reached into the cab and brought out half a dozen letters. One was for me from Liz.

The strain of writing daily had taken its toll fairly early. We'd been out six months and this was her fourth letter. The envelope smelled of perfume and I sniffed it a couple of times before ripping it open.

No wonder she hadn't written. Her parents had whisked her off to Europe for six weeks. Had I received the post cards? When would I be back? Tippy's going to be a mum. I told Mowie. He blinked stoically a couple of times, then went back to chewing a rotting bone.

Milton, one of the shearers, also received a letter from his wife — telling him she'd left him for some smooth schoolteacher who lived in Adelaide and had swept her off her feet.

"I wonder how smooth he'll look without any teeth," Milton muttered.

Nothing could have made the shed seem so hopelessly remote as this news.

Milton said he couldn't bear to think about it — but he often did. We were still weeks away from finishing the run and he tossed up the possibilities of racing home for a weekend to sort things out.

On Saturday night he became morose and drank himself into a talkative stupor.

He was a pathetic sight, dressed in scruffy shorts and an old army issue shirt, facing an evening with a rowdy mob of shearers and shedhands.

"She'll be goin' out tonight for sure," he murmured. "She's beautiful. Long hair. Long legs. Black stockings. Bastard. Bastard!"

His eyes blazed, then softened and filled with tears. "I'd take her back." Just a whisper.

Milton read and re-read his wife's short letter, hoping to find some hidden ray of hope somewhere between the lines. There wasn't any. The episode was simply a symptom of the strain that shed life puts on relationships.

Like sailors, shearing teams are away from home more than they're at home. But whereas most sailors have visits to exotic ports and night life at least once every few weeks to break the monotony, shearing teams don't.

High spots in a long dreary run might be nothing more than visits to yet another bush pub. Male company, and rough male company at that. The pleasures and comforts of home life seem light years away.

In some cases absence might make the heart grow fonder. In many cases the heart hardens and eventually gets to the point where it couldn't care less.

Poor communciations don't help. Infrequent mails. Non-existent telephones. They both take their toll.

It must have been something to do with the weather or something. First Mowie. Then Milton. And then me!

Liz wrote one more letter. Short and to the point. I blinked stoically too, but I turned down Mowie's offer to share a bone.

After cut-out I drove almost all the way to Adelaide before I really thought about it and decided that I didn't even have to go back to Adelaide.

The run was over. I was probably faced with a couple of months' spell. I had some money. Why not do something different? Go somewhere different?

Western Australia seemed like a good idea. Now might be as good a chance as any. Why not, indeed?

In early morning sunlight I pulled up by a phone box at Gepp's Cross. Mowie sat up on the front seat and watched me as I crunched across the gravel verge and pulled the door open. My parents were still in bed.

Mum's sleep-croaked voice said "Hello?" I told her what had happened and what I'd decided to do.

"Western Australia? Why Western Australia, son?"

"I've always wanted to work there. I'll be back in three or four months."

Mum's voice strengthened as sleep retreated. She sounded concerned but she didn't try to change my mind.

"Say goodbye to your father."

She passed the phone to him. He was still half asleep but he wished me well.

"Here's your Mum again, lad."

Mum's voice came back on the line. "Take care then. God Bless."

"See you."

Mowie was still sitting watching as I strode over to a roadside rubbish bin. Liz's last letter landed on top of an ant-ridden chicken carcase. I wheeled the van around and headed back the way we'd come.

An uneasy mix of anticipation and apprehension nagged at me.

I had very few contacts in the West. Just a couple of skindiving mates I'd met during a holiday in Perth.

As the road drummed under the wheels I realised that my parents, along with most other migrants, had left their homes for an entirely new country without any guarantees of survival.

Some of those migrants couldn't even speak English when they arrived. At least I could speak Western Australian.

Western Australia didn't seem so far away then. Apprehension eased. Anticipation climbed.

Somewhere west of Coolgardie the night sky was seared by enormous forks of lightning. Kilometre after kilometre of lightly wooded country was illuminated by each flash.

I'd been driving almost non-stop. It was close to midnight so I decided to take a break.

At the bottom of a low hill a track swung away to the north. I wheeled off the highway, followed the track for a kilometre or so and switched off.

Heavy thunder boomed across the sky. Rather than risk being caught out in the rain I boiled the billy on my little metho stove inside the van.

Mowie either didn't know that it was going to rain, or he didn't care. He bounded over the open tailgate and raced along the fenceline, sniffing and piddling against almost every post.

Within a few minutes he had all but disappeared and only the bobbing white tip of his tail showed me where he was. I called him back and he came hurtling in over the tailgate, almost knocking my fresh brew over.

There were a few butcher's scraps left in the esky so I gave him those to chew and he settled in a corner to make a meal out of them.

As I sipped my tea the first big raindrops splattered on to the roof of the van. A solid thunderclap announced the downpour and, seconds later, rain was hissing down in sheets. I slammed the tailgate shut. Snug and dry, I swilled the last of the tea, patted Mowie and turned in.

Streams of water sluiced down over the windows. Gusty winds rocked me to sleep.

Mowie's whingeing roused me. It was five-thirty and only just light. Heavy overcast. Stiff joints. The smell of rain-washed earth.

Mowie scrambled out of the van when I opened the tailgate. Some soggy sheep had drifted up the fenceline during the night but the sight of Mowie sent them scurrying into the scrub. I called him back, just in case, but he was too busy cocking his leg against the fenceposts to worry about chasing sheep.

Breakfast for both of us was cornflakes without milk.

Then back down the muddy track, past clumps of smooth, copper-barked trees, and on to Perth.

6
West of Centre

WISE MEN FROM THE EAST have never been popular in Western Australia. Too many of them have moved across with the intention of showing the locals how to do it. This, and isolation from the rest of Australia, has led to a strong parochialism in the West which is often hard to penetrate.

Despite a ream of good wool reports and a few glowing references, I couldn't get a job as a shed classer to save my life.

I'd been able to move in with the Butlers, a mate's family, in Cottesloe, a suburb of Perth. It was just like home. Every morning Mrs B would make me a cup of tea and give me first look at the morning paper, *The West Australian*. I'd scour the "situations vacant" columns and ring for every woolclassing position advertised. The story was nearly always the same.

"Are you experienced?"

"Yes."

"Where've you been classing?"

"South Australia, Victoria, New South Wales."

"Ahh . . . er, we're looking for a local bloke. Thanks for calling, though."

My knowledge of Western Australian geography wasn't sound enough to let me lie about where I'd been, so I was stuck. But eventually I got a job in the woolstores at Spearwood.

Only impending poverty forced me to stay. There was none of the camaraderie I'd known in Port Adelaide. The pay was dismal, the work stupefying. But it was that or nothing. Anti-eastern states feeling was even more noticeable there. I was given a job sorting through oddments — fragments of fleeces, crutchings, bellies. Menial work, but, at first, welcome enough.

Then it turned out that I was the only classer in the section who had any shed experience. All the local blokes were stores' trained and that's where they would stay. Most of them hardly spoke to me. Seeing those store classers being given bales of fleece wool to class, while I grubbed and picked my way through oddments day after day, started to rankle. I complained.

The foreman told me I'd get fleece wool to class when I proved myself. I offered to show him my stencil and wool reports. He didn't want to know.

Weeks went by.

Oddments are time-consuming and my tally of bales was way below those of the classers who were working on the fleece lines. The supervisor complained to the foreman.

I'd only just arrived for work one morning when the foreman confronted me to pass on the supervisor's message.

My rolled-up apron hit the floor between his feet and I was out of a job again.

But this time fortune smiled. Next day there was an ad for a rouseabout in a shed.

"Are you experienced?"

"Yes!"

"Where've you been working?"

"I've been classing in . . . er . . . Gmmmddggggrrrrup". That sounded like a West Australian name. And it worked!

"Oh, good. We're looking for a classer as well."

"Do you need a dog too . . .?"

Things were quite different at sheds in the West. The first jolt came when I arrived at the shed near Dowerin in the wheatbelt.

"Where are the woolclasser's quarters?" I asked.

"Wherever there's a spare bed," the contractor replied.

Different indeed from the other States, where the classer is always housed separately, and the property owner supplies the bedding.

I went looking for a spare bed — and found one. My roommate's bed was made very neatly but there was no sign of him.

"You can't stay there, woolclasser!" It was the contractor again.
"Why not?"
"That's Beryl's room."
"Female cook?"
"Na. Female rouseabout."

I almost fell over. I'd read about female shedworkers in New Zealand, but I'd never heard of them in Australia.

"What's she like?"
"Good worker."

In fact, she was hopeless. She couldn't throw a fleece. She wouldn't sweep up. She was slow. And she was usually two or three minutes late for the start of each run. She had no idea about wool, so I often found bellies in the locks, and pieces in the bellies.

But she did have a pretty face and a great body. And that was the biggest problem. Half the team were in love with her. She wore denim shorts that looked as though they'd been painted on to her thighs and a deep-necked blouse loosely knotted above her navel. Each time she bent over to pick up a fleece, every eye in the shed went with her. Whenever I tried to hurry her up, one or more of the shearers would step in and defend her.

Petty jealousies sprang up between the blokes as they all tried to win her. Some of the blokes succeeded. One night someone sprinkled talcum powder outside her door. In the morning, the powder showed the unmistakable tracks of a barefooted visitor.

Next night, around midnight, when the quarters were reverberating to the sound of heavy snoring, a muffled but loud, "*aaaaaaaaagh . . . ssssshhhhhhiiiiiittt! I'm crippled!*" woke everybody.

Torch beams shone down the passage way. Larry was quick to hobble back into his room — but not quick enough to avoid being seen, though. He'd been hopping round in pain near Beryl's door. On the floor were a dozen or so drawing-pins.

By the time cut-out came three weeks later, I'd had enough. I confronted the contractor.

"Either Beryl goes, or I go."

No hesitation about the reply.

"I'll make your cheque out right away. Do you want to sell your dog?"

Back to reading the situations vacant columns in *The West Australian*. One ad caught my eye: *Woolclasser wanted. Wages or contract. Long run up north. Good sheds. Must be top man.*

That sounded like me! I rang straight away and organised an interview.

Peter Margate lived in a comfortable house in the suburb of Innaloo. He didn't bother to get up to answer the door. He just yelled "Come in!"

He was stretched out on the sofa watching the cricket and listening to the races simultaneously. There were half a dozen empty cans of Swan lager lying at the foot of the sofa.

When they'd passed the post at Randwick he looked up and spoke to me.

"What I want is a bloke who can run the shed when I'm not there."

"I can do that. Been doing it for ages."

"Classer I had last year was a good bloke, but he went and got married. Now his wife doesn't want him to go away. He's gone back to the stores. Can't understand that myself. You're not married, are you?"

"Nope."

"Right, it's yours. We start at Meekatharra on Monday week. Don't suppose you've got a good dog, have you?"

COOLA DOWNS

ONE HUNDRED kilometres out from Meekatharra, Coola Downs couldn't have been hotter.

Heat haze shimmered off the airstrip that stretched alongside the shed and quarters. The strip was really an extension of the track that led into Coola, straightened and widened like the hammered-flat end of a length of kinked fencing wire. White-painted tyres down each side marked the boundary of the strip.

I followed Peter Margate's car as he drove down one side of the strip, then swung across it to pull up near the quarters. Peter drove the latest model Ford Galaxy. Swish, air-conditioned luxury. We crunched to a stop outside the quarters.

Everybody else had already arrived and were sitting in the shade of the mess hut, swigging on cans of Emu Export.

They were a mixed mob of blokes. One shearer from Tasmania, Max. Two from South Australia, Ferret and Dinky. One from New Zealand, Herbert. Two from Western Australia, Geoff and Eric.

Eric was one of Peter's old guards. He had been with him for five or six years. One of the woolrollers came from NSW, the other from South Australia — Martin and David. The presser, board-rouseabouts and cook were all from Western Australia, Malcolm, Jimmy, Norm and Cec.

Peter himself only intended to stay a couple of days in the shed. He'd spent years in the bush and was more interested in the diversions of city life these days.

I was classing and overseeing. In practice that only meant classing, counting out the sheep and signing the cheques to pay the blokes. The shearers were going to do their own grinding. Eric would do the experting. The presser would pen the sheep. It was a holiday shed almost. Just the heat, dust and isolation to put up with.

Herbert was part-Maori. A big, solid bloke with log-like arms and tree-trunk legs. I thought it might have only been the size of the Kiwi that stopped the other blokes from throwing him out of the shed the next morning. He had "pulled" combs. Combs with their teeth stretched wider apart to take in a wider swath of wool on each blow.

"Bloody hell, Peter. There's going to be bloodshed here!" I said.

"Naaaaa! Don't be stupid. This is Western Australia. Those blokes won't worry about pulled combs here."

Anywhere else in Australia, pulled combs, or the manufactured wide combs that New Zealanders use on their open-wooled sheep, were as popular as pork chops in synagogues. Union rules dictated that every man must have the same standard combs to give everyone a fair go. Anything other than the standard comb was forbidden even to enter a shed.

But the union didn't have a strong voice in Western Australia at that time. In this team, at least, nobody objected to Herbert's pulled combs.

Coola Downs turned out to be an interesting shed. For two reasons. Even though Herbert had pulled combs, he didn't shear the most sheep.

Max, the Tasmanian, did that, shearing with standard-width combs. Herbert came second. His sheep were beautifully shorn, but he always used one or two less combs every day than the other blokes.

"How come, Herbert?" someone asked him.

"Ohh, I dunno. I used sucksteen today. Ow many did yous use?"

"Eighteen!"

"Eighteen!"

"Nineteen!"

Herbert shrugged his great shoulders. "Must be in the grinding, I s'pose."

So we watched him grind his combs, eagerly waiting to be shown the delicate touch of a master grinder. His great meaty hands, their backs covered in thick black bristles, jammed a comb against the disc. Sparks flew.

Gggggggggrrrrrrrr! The diesel motor staggered under the pressure.

Herbert finished up with one long push. Pressed hard against the disc. The disc rocked on its pedestal. The tips of the comb glowed hot.

Struth !

But that wasn't all. When he'd given the comb that finishing touch, it had been hard up against the central nut that holds the disc in place.

Fifteen minutes later, when the comb was cool enough to pick up without swaddling it in a wet sack, I looked at the grinding marks on it. They ran across the teeth at right angles.

Every other shearer I've ever met insists that the grinding finish near the rim of the disc so the grinding marks run up and down, parallel with the teeth. Perhaps Herbert had discovered something new.

☆ ☆ ☆

We had a long drive ahead of us when we finished Coola Downs. The next shed was 300 kilometres away, just the other side of the microscopic town of Agnew.

Agony, as the locals called it, was once the centre of a thriving goldfield but it had suffered the same fate as countless other gold towns. The gold had paid out. All that remained during our visit were a few dilapidated houses inhabited by diehards and a combined store, post office and pub.

It was Friday afternoon. The entire team pulled up at the pub to sluice the dust from their throats. There was only one other drinker in the bar when we walked in — and he didn't look like a local. He was neatly dressed in shorts and long socks and he'd had a shave recently. Unmistakably a city-ite.

"Come on, woolclasser, it's about time you bought us a beer, you miserable eastern states' bastard." That was Malcolm, the presser.

As I fished for enough money to buy the round, the city-ite grinned and said, "Don't argue mate. There're too many of 'em."

His glass was empty too, so I offered to fill it.

"Thanks. Cheers, Jim's my name and I'm an eastern states' bastard too." He didn't realise it then, but most of the blokes in the bar fitted the same description. Jim was a car salesman, making this trip into tiger country to try to sell an expensive car to one of the richer station owners.

He'd moved to the west from his native Canberra because he'd heard that the fishing was pretty good.

"Not many fish around here, Jim."

"No, but I'm going to make up for it when I get back. I'm going to buy a boat." We yarned on and bought each other a few beers.

Around us the team became more boisterous. Someone started to sing. Two of the blokes waltzed each other round the bar. That didn't take long because the Agnew pub has one of the smallest bars you could imagine, but they kept going till they got dizzy and collapsed in a heap.

By the time darkness fell, most of them were too drunk to stand. Jim had become Lloydy. I'd become Hewsey.

A keg's dying gasp signalled closing time and those of the team that could walk helped carry out those who couldn't. They sorted themselves out into carloads and set out on their rather hazardous way to the next shed.

Roaring motors and crunching gears pointed to the cars that were being driven by unfamiliar drivers, standing in for team-mates incapable of driving.

Lloydy, silhouetted against the yellow light from the bar, waved me off.

"See you in Perth. We'll go out in the boat."

I drove for a while with the window open and my head out in the slipstream to keep myself awake. It was a freezing, clear night. Overhead the Milky Way blazed in a black-velvet sky.

"You know what, Mo? I wouldn't give up the bush for quids."
Mowie slept on.

DINKY, GEORGE AND FERRET

"SHE'S GOT SWALLOWS tattooed on her tits!" Dinky sounded incredulous. He was standing in my room, waiting while I wrote a cheque out for him.

It was Friday night. We were 250 kilometres from Kalgoorlie but Dinky and Ferret were going into town for the weekend. They were going to travel with a couple of Kalgoorlie prostitutes.

The girls were an enterprising pair. They had driven all about the countryside, visiting mining camps, shearing sheds and lonely prospectors. They'd arrived at our shed just after tea.

Dinky hadn't wasted any time in getting to know them. Now they'd talked him and Ferret into travelling with them back to Kalgoorlie.

"How are you going to get back?"

"Oh, they'll drop us back on Sund'y night on their way out again."

"They'd better."

He grabbed his cheque.

The Holden wagon roared into the night — Ferret in the front with the driver, Dinky and his tattooed companion locked in an embrace on the back seat. "Bet they'll feel every bump," somebody drawled.

Sunday night came without any sign of Dinky and Ferret.

But on Monday morning, there was a strange car parked out in front of the quarters. As I watched, a hand started to wipe away some of the condensation from inside the front window. A bleary face and a thatch of tangled silver hair appeared, framed in the smeary-clear window.

The driver's side front door opened and an old bloke, wrapped in a holey army blanket, struggled out from behind the wheel. He looked bewildered.

Then Ferret appeared from his room and sheepishly walked up

to the car. "Er g'day woolclasser . . . g'day George."

George was confused.

"G'day Ferret . . . er where the friggin' hell is this?"

"Er . . . Milboona."

"What? . . . what? I'm supposed to be in Leonora."

Ferret looked very uncomfortable. "Yairs . . . er . . . well, we thought yer'd wake up soon enough to drive out there this mornin'. That's why we left yer in the car . . . er . . . yer was too drunk to drive there last night George." It was patently clear George had been kidnapped for his car.

By late Sunday afternoon Dinky and Ferret had been dumped by their escorts — and they were getting desperate for a ride back to the shed. They did the rounds of the pubs, looking for someone who was heading our way. They'd spotted George.

George had been drinking alone so they lost no time buying him a round or two. He was due to start shearing at Leonora next morning. That was in our general direction except he declined the chance to take a sizeable detour to drop the desperate duo back at our shed.

But by the time the session finished Dinky and the rouseabout had George stewed to the eyeballs. They carried him out of the pub, propped him up in the back seat of his own car and then drove off with him. That would cost George his job.

He was known to be an alcoholic — although he was a good shearer when he was sober.

When he failed to arrive at the Leonora shed, he'd be scrubbed off and his pen earmarked for someone else. George would have been justified in wanting to kill Dinky and Ferret. But he just laughed it off. "Can't blame the young blokes."

I talked it over with the rest of the team and they agreed to let me set George up on a spare stand at the end of the board. There were a few raised eyebrows when I finished counting out at the end of the day.

George's tally seemed remarkably high for someone suffering from an acute hangover. But for some reason that only the gods of retribution might understand, Dinky's and Ferret's tallies had dropped by twenty-five or so.

☆ ☆ ☆

Six shearers going flat-strap. Wool everywhere. Sweat running in rivers.

"Black wool!" one shearer shouted, and David raced up the board to answer the call.

Eleven-thirty. Half an hour to lunch. A chance to get the shed tidied up.

Then Geoff, one of the local shearers, strode down the board with both a fleece and David struggling in his arms. Geoff dumped them unceremoniously on the classing table in a heap.

"What the hell are you doing, Geoff?" I asked.

"He was slow picking up the fleece!"

David rolled off the table with a big grin on his face and went to move back down the board.

"Hey, straighten that bloody fleece out." The voice of authority was definitely mine — or I thought it was!

"Tell him to get stuffed," Geoff told David, referring to me.

I was furious but Jimmy and Malcolm thought it was a big joke. So did most of the shearers. Maintaining discipline in a shed is difficult enough in a loosely governed group like a shearing team. This kind of behaviour made it almost impossible.

I called Eric. He was Peter's right-hand man. "I'm going to fire Geoff, mate!"

"Aw, come on, Hewsey. He's only fooling around."

"I mean it, Eric."

"All right, I'll see if he'll apologise first, okay?"

Geoff said: "Tell him to get stuffed."

He left that night, and Eric didn't speak to me for three days.

FERRET DROPS OUT

WE SHORE ON till cut-out then had another long drive out from Agnew, through Meekatharra, then north-west through Gascoyne Junction to another shed near the coast, north of Carnarvon. My van was following in the wake of Jimmy's Mazda.

Thirty kilometres out from Meekatharra, disaster struck. Jimmy's Mazda lurched drunkenly across the road, hit a big rock with the rear wheel and swerved back on to the road in a cloud of dust. Undaunted, Jimmy kept going. Not so Ferret. He had been comatose in the back seat. A long, hard day in the Meekatharra pub had anaesthetised him for the arduous drive out to the shed. The big rock had flexed the frame of Jimmy's rusting Mazda and the back door had popped open. That was the door Ferret had been leaning against. He fell out. At about eighty kilometres an hour.

When I pulled up Ferret was sitting near the graded edge of the road, swaying slightly and murmuring: "Strike me pink...strike me bloody pink."

As far as I could tell, he still had the skin under his armpits. Most of the rest was smeared over the few metres of road he'd covered as he skidded to a stop. Along with the skin was a considerable percentage of his best "going-out" footy shorts, nearly all his blue singlet and one thong. He didn't look well.

A glance up the road at a diminishing dust cloud confirmed that, as yet, Jimmy hadn't noticed Ferret's absence!

There was still life in Ferret's battered form but I hesitated to risk moving him in case he'd broken something. It was hard to believe that he hadn't broken everything. His weeping injuries attracted flies in their hordes, but he didn't seem to notice them.

I was tossing up whether to stay with him or leave him and go for help, when the buzzing of the blowflies was drowned by the drone of Jimmy's returning Mazda. Pluming dust reared up behind the car as it crested the last hill and then roared downhill towards us.

From the mobile wreck stepped an alien form red from head to foot. Even the half-full beer bottle, clutched in one hand, was wearing a coating of red dust. Bloodshot eyes blinked in amazement. Using verbal shorthand, Jimmy explained that he had been unable to work out why his car was taking in so much dust.

When he'd turned to ask Ferret's opinion, he'd discovered that not only had Ferret got out, he'd also forgotten to close the door after him. "Look at the bloody mess!" he finished.

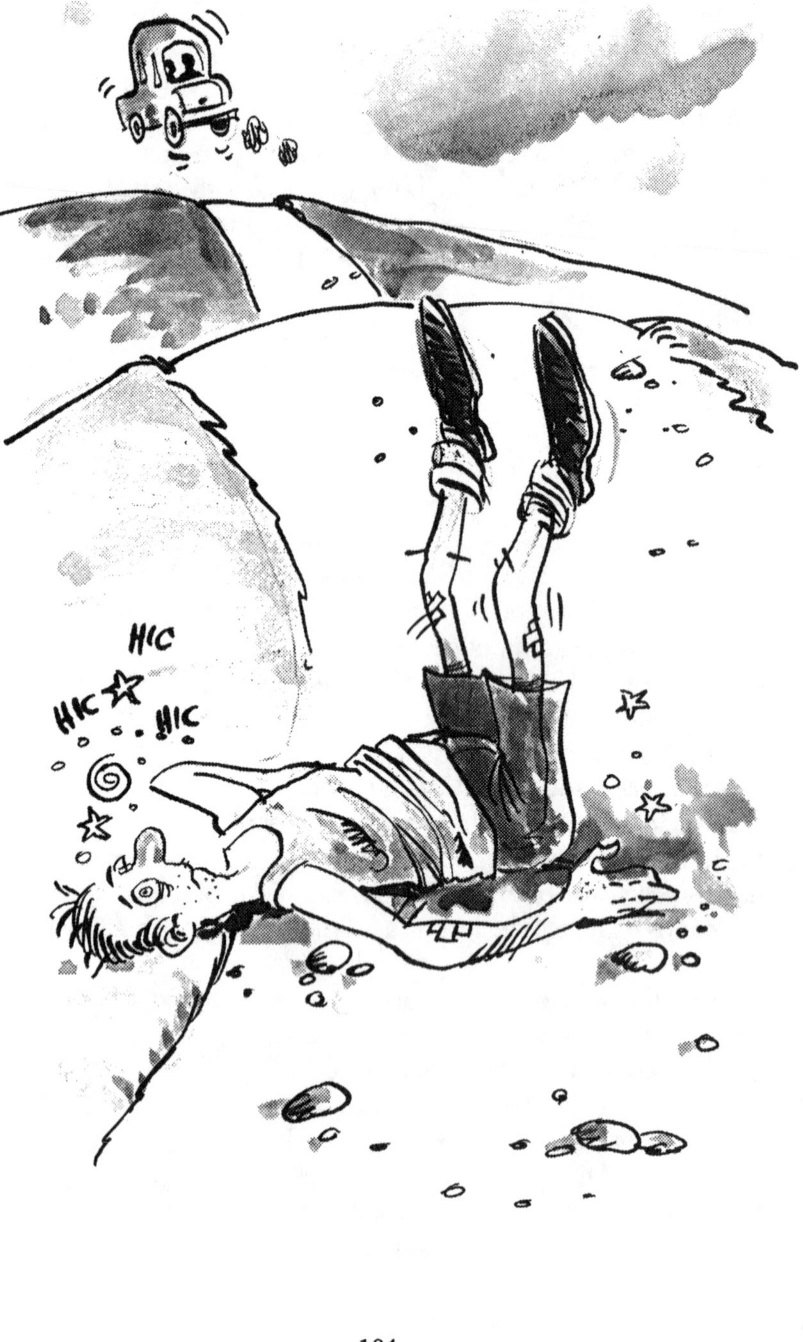
HIC
HIC
HIC

I wasn't sure if he meant the interior of his car or the exterior of Ferret.

Up to this point, Ferret's sole contribution to the conversation continued to be: "Strike me pink . . . strike me bloody pink," but he now added a few groans. He was also tentatively probing his sorer parts with a shaky finger.

"Can you wiggle your toes, Ferret?" I asked. Ferret nodded his head but made no other move. I persisted. "Ferret, move your toes, mate, we need to see if you've broken your legs."

This time, slowly and painfully, Ferret wiggled first the toes on the foot that still held a thong, then the others.

Having plumbed the depths of my first-aid knowledge, I thought it best to get Ferret back to Meekatharra as quickly as possible.

"Can you stand up, mate?"

Ferret nodded his head dazedly but made no other move. It was apparent that we'd have to pick him up. Jimmy was not a great help, being almost as unsteady on his feet as the dented Ferret, but we grabbed a hand each and levered him into as upright a position as he could manage. No doubt the effect of the alcohol helped take some of the sting out of his movements, but poor Ferret was clearly in great pain.

We cleared a space in the back of my van and eased Ferret into it after padding the floor with my bedroll and covering the rough army-issue blankets with a couple of cleanish towels. He seemed a pathetic little heap as he lay there.

Then with Ferret semi-comfortable in the back and Mowie snoring on the front seat, I wheeled the van round and headed back to Meekatharra, 120 kilometres over rough dirt road. Ferret felt every bump.

Steadfastly sucking on a fresh bottle, and far enough back to avoid our dust, Jimmy followed us "to make sure 'e'sorrite".

Clear, cold darkness seeped through the bush as night fell. Ferret's groans grew more frequent as his body woke up to what had happened to it. After countless bumps and jolts, a few anaemic lights appeared in the near distance. A few minutes later the bumps disappeared as smooth bitumen rolled under the wheels. We were back in Meekatharra.

Fortunately the local doctor was home. In pyjamas and dressing-gown, he peered into the back of the van, made a cursory examination of Ferret, then told us to carry him into the surgery.

Ferret's immediate problems were over — but mine were just beginning. We were now a shearer short to start work on Monday morning. An hour remained before official closing time, so I made a round of all the pubs to see if I could raise someone to take Ferret's place. There were quite a few shearers in town that night, but they were all employed. No luck.

Jimmy meanwhile stayed on at the Railway Hotel for a nightcap.

I knew there would be little chance of finding Peter Margate at home in Perth — it was Saturday night — but I swatted my way into the moth-infested phone box and asked the operator to try his number. Several minutes later I was surprised to be put through to Peter.

He had a party going full swing in the Innaloo house: it had obviously been going on for most of the day. His speech was a fair imitation of a 45 rpm record being played at $33\frac{1}{3}$rd. After assuring me several times that there were "no worries, mate" he asked me over for a drink, then hung up. Or fell over.

Paralytic moths swooped and fluttered round me as I gathered up my unused coins. The whole bloody world seemed to be populated with drunks. I pulled a partly chewed moth from the side of Mowie's mouth, bundled him back into the van and roared off up the road back to the Railway. I wasn't in the best of humour.

Rooms in outback pubs are not flash; the Meekatharra pub was no exception. I was tired, though, and willingly tumbled into a hard, but clean, bed. Mowie was chained to the roo bar on the van and God only knew where Jimmy was. I didn't care. Soon I had slipped off to the land of Nod.

☆ ☆ ☆

Lemony sunlight lanced through the thin lace curtains and gently, but persistently, woke me up just as a low-flying aircraft roared overhead. Although I didn't know it, that was the Flying Doctor coming to collect Ferret and transfer him to Perth.

I hunted out some breakfast in the kitchen, and the old cook was good enough to unearth the remains of a roast shoulder of lamb for Mowie. This was gratefully received; I left him soaking up the first warmth of the new day with his sleek black coat and avidly tearing into his breakfast.

Then it was my turn. Flash five-star pubs in the capital cities cannot produce food with the character or flavour of bush cooked breakfasts. Perhaps it's because the city pubs use clean pans or fresh cooking oil. Bacon, eggs and sausages cooked in the pungent, smoky atmosphere of a wood stove and swilled down with hot, strong tea leave a man feeling well fed and ready to face a good day's work.

First task was to locate Jimmy. With a bit of luck it should be possible to reach the next shed today and get it underway on the morrow.

Jimmy wasn't difficult to flnd. After his nightcaps, insulated from the cold by alcohol, at least temporarily, he'd slithered into a heap on the dusty back seat of his car and spent an uncomfortable night parked behind the pub. When I came across him, he was having a smoke and trying to get some feeling back into his legs. He was pretty seedy.

A quick trip to the surgery told us of Ferret's impending departure. When we saw him he tried to smile but it hurt too much.

No wonder he couldn't stand straight when we tried to move him after the accident! As he bounced and scraped his way along the road he'd broken some ribs and delivered a mighty thump to his coccycx. The doc was concerned about his spine.

We stood and watched him wheeled out on his way to the plane, then took to the road. I sent Jimmy on ahead and made a brief stop at the phone box. I knew Peter wouldn't be in a fit state to talk, so I stuffed a heap of coins in the slot and sent a telegram. *FERRET CROOK. SEND REPLACEMENT. HUGHES.*

Jimmy was out of sight by the time I swung out on to the road.

7

Beach of a Life

THE PRESSURE WENT ON in the shed near the coast. Eight shearers. Good sheep. Plenty of wool.

There were three new shearers: Merv to replace the injured Ferret, Russ and Bill to take up the extra stands. And a full-time penner-up: Graham.

Graham was a liability. He'd approached Peter Margate for a job because the demons were chasing him for overdue maintenance payments. Since Margate spent half his life in the same predicament, he showed Graham some uncharacteristic sympathy and sent him up to us. Penning-up requires commonsense and perseverance. Graham didn't have much of either.

Mowie helped him out a little. Half the time, though, the dog was just as confused as the sheep. Disjointed orders had him trying to "go back, ya bastard" when he'd already gone as far back as he could without jumping out of a shed window. Or trying to "lie down . . . come 'ere, ya mongrel" at one and the same time.

Shearers were often left with empty pens and Graham took lots of stick because of that. But he just shouted louder abuse at Mowie and kept struggling along.

Sometimes Malcolm, the presser, myself, or, on a few occasions, the shearers themselves, jumped over the rails and helped move sheep into a catching pen so that they could keep shearing. That sort of thing quickly starts friction in a shearing team. Graham wasn't popular at all. He was putting everybody else under more pressure than they needed to be.

On the other hand, our weekends were marvellous. Few sheds

have the luxury of a tropical beach within a short drive of the quarters. This one did: Coral Bay.

We spent our weekends lazing on the beach, or swimming in gin-clear tropical water, with an air-conditioned pub on hand to serve icy beer and counter lunches.

Coral Bay was originally developed as a millionaires' playground, intended to rival the lush resorts of tropical Queensland.

However, the developers overlooked the fact that any resemblance to North Queensland ends with the pristine sand and clear water. Coral Bay sits on a desert coast, with no waving palms to fan the romantic aspirations of visiting couples.

The pub itself offered the only available shade, and it was usually frequented by the likes of us: shearing teams, roadgangs, professional fishermen.

The facilities weren't great, either. Salt-water showers. Toilets that frequently broke down. The swimming pool usually empty of water but half full of beer cans. Although now vastly improved, Coral Bay in those days was hardly the kind of place you'd be likely to trip over the Aga Khan!

Lack of free-spending international visitors soon squeezed the infant resort into a tight corner, but for those used to spending their weekends in the confines of parched, dusty, isolated shearing quarters, Coral Bay was akin to paradise. Weekends there helped reweld the team together.

But Graham over-indulged. Considering his financial predicament, he seemed to be quite unconcerned at the proportion of his pay cheque that was disappearing down the pub's urinal. He was fully drawn up and became snaky when I refused any advances on his wages. He was usually so crook on Monday mornings that he bordered on the useless anyway, but thoughts of his wife and kids waiting hopefully for maintenance prevented me from sacking him. Still, he was living on borrowed time.

Mowie was doing his fair share of the penning-up and most of Graham's as well, so I blew up when I heard my dog yelp in pain.

"Stuffin' dog's bloody useless!" was Graham's feeble excuse as I charged into the pen to confront him. He was a wretched sight when faced with a crisis.

I vented my anger by ordering Mo out of the shed. Poor blighter, he must have wondered what on earth he'd done to deserve all the attention. Without Mo's assistance, Graham discovered just who was "bloody useless" and it was only a half-hour or so before the shearers were bellowing because they were running short of sheep. I let him stew in his own juice.

The following weekend saw one of Graham's usual binges but he received a bigger surprise than the rest of us when Peter turned up unannounced on Sunday night. He'd tired of the city, or a least decided that he needed some respite from the strain the city placed on his bank balance, and had come up from Perth to inspect his woolly, sweaty empire.

Although Peter was no teetotaller himself, he viewed Graham's condition with disdain.

"You said you were going on the wagon until you'd paid your old lady off."

"Stuff her!"

"What about your bloody kids?"

No answer.

By the end of the first run on Monday morning, Peter had seen enough to convince him that we were carrying a lame duck. "I'll toss yer to see who tramps 'im," he said to me.

It must have been one of those coins with two heads because I was soon on my way to the yards outside the shed to rid the team of Graham.

He didn't look at all surprised when I told him. He just calmly squared up and heaved a haymaker at me. It missed, but evasive action overbalanced me and set me down in the dust. Peter, peering through one of the portholes, roared with laughter.

Graham avoided retaliation by leaping the fence, then, cool as you like, he turned and said: "Well, how much loot have I got in the kitty, then?"

It didn't take long to work that out. A day's pay less some stores and ten dollars he owed the presser left Graham with about enough for a tram fare.

His attempts to raise further loans from the team members fell on stony ground and he resorted to offering various items of clothes to the rouseabouts at bargain basement prices. That, too, was unsuccessful. Finally, in a fit of pity, Peter stuffed ten dollars into his hand and drove him to the Coral Bay turn-off to wait for a truck heading southwards.

Any man with an ounce of nous would have waited there and climbed into the first vehicle to offer itself. Not Graham.

He picked up a ride into Coral Bay and made himself comfortable at the bar. Ten dollars doesn't go very far in an outback pub and Graham was soon broke again, but he had his story ready and found a pair of sympathetic ears to listen to it.

The barman's wife at Coral Bay sometimes served behind the bar while her husband busied himself with other jobs. Graham told her his story.

"Yeah, right bastard it was," he began.

"How's that?"

"Well, I've been shearin' two 'undred a day. Ringin' the shed. Makin' a big quid . . ."

"You must be a real gun . . ."

"Well, I don't talk about it much. Anyway, there I was. Twenty ahead for the run when this dirty big wether pulled me over and strained me back."

"Oh, no!"

"Bloody agony!"

"I'll bet."

"Couldn't straighten up."

"No wonder."

"So I've got to go see a specialist in Perth."

"Sooner the better, I'd say."

"The boss's been pretty good about it, though. Dropped me off here and told me to make meself comfortable until he could come in on the weekend and drive me down to Perth 'imself."

"Struth! That's good of him!"

"Yeah, 'e didn't want me doin' more damage by bouncin' round in the mail-truck."

"That *is* good of him."

"Yeah, 'e's got to get another shearer in t'take me place and there ain't any room for an extra bloke out at the shed, so 'e's even payin' for me to stop 'ere. 'E's a good bloke awright. Just put it on the tab and I'll fix it up on Sat'dy' — that's what 'e said."

"Struth. Wish I had a boss like that... do you want another beer, love?"

"Make it a scotch. Johnny Walker. Double."

"Righto, love. Then I'll show you to your room."

There was no telephone out to the shed, so there was no way that Graham's story could be verified, and for the rest of that week he lived in air-conditioned luxury. He became a regular in the à la carte dining-room and modified his drinking habits to accommodate regular scotches instead of beer.

During the week, it was usually ten o'clock before he appeared for breakfast. But on Saturday morning he slipped out of his room before first light, walked a few kilometres back towards the turn-off, and caught a truck heading south.

About lunch time Peter breasted the bar at the pub; along with his first drink he received the bill for Graham's stay. He refused to pay. The barman emerged from the bar and proposed making some major modification to Peter's face if he didn't come good with the cash, whereupon Peter responded with some dire threats of his own, and in a matter of seconds they were at it like two stray cats.

Peter was pretty good with his front feet and he got in a few resounding clouts to the side of the barman's head before he copped one on the nose. That enraged him and he charged forwards, carrying the barman before him out of the door and on to the patio outside. The barman's wife ran after them shrieking and swearing — an activity which she only gave up when the struggling men barged into her and knocked her into the waterless pool.

By that time the entire population of Coral Bay, which you could count on two hands, had gathered on the patio to watch the fight, by now looking decidedly gladiatorial.

The barman had picked up a table and, holding it by the top, he was trying at one and the same time to ward Peter off and club him to death with its legs.

Peter in turn had selected the shade umbrella from another

table and was trying to insert it in the hole in the centre of the table being brandished by the barman. Had he succeeded there's no doubt that the pointed, galvanised tip would have skewered his opponent.

However, the hole proved too elusive a target and despairing of success but still very angry, Peter hurled the brolly away and seized one of the table legs. There was a feverish tug-of-war.

Finally Peter saw his chance and swung one of his large fists in an arc over the intervening table. He connected and the barman reeled, dropping the table as he did so. Peter followed up with two solid clouts to the jaw and the barman collapsed.

Moments later the hotel manager appeared and the whole team was banned from entering the pub ever again.

That put a bit of a dampener on weekends. We still went to Coral Bay, but instead of sipping beer in the air-conditioned bar, we sat in whatever shade we could find outside and relied on members of a road-gang to ferry us cold drinks.

For the next few weeks, hangovers were displaced as the number-one Monday morning malady. What took their place? Sunburn.

MAX DOES HIS SCONE

AFTER OUR STINT in paradise it was back into tiger country. Towards Gascoyne Junction to start with, then across country, back through Meekatharra and out towards Sandstone. A bit of a shock to the system. Certainly too much of a shock for Cec, the cook.

"Hey, woolclasser, wake up! The cook's snatched the rent!"

One of the shearers, out of bed for an early morning wee, had discovered that the cook had disappeared overnight.

It took a while for the news to penetrate. At six o'clock on a frosty morning, sleep, beneath a mound of blankets, is not easily surrendered.

Cec had always had a problem: his scones weren't light enough and everybody had rubbished him over that. It seemed there had been one good-natured insult too many. He was almost fully drawn up — he'd said that he needed to pay off a debt, so I'd had no hesitation in giving him the money. But now? What about breakfast?

Grudgingly I left my bed, gasped as my feet hit the icy floor and struggled into my clothes.

It was Friday; the least Cec could have done was to hang about till the weekend. This would have given us time to find a replacement, probably from Carnarvon. As it was we looked like having to face a working day without a cooked breakfast. And what about smokos? And lunch?

Max Jackson, our gun shearer, came to the rescue. He offered to cook breakfast and added that he'd knock up a batch of scones for smoko, too. It turned out that he'd been a cook at a woodcutters' camp in Tasmania at some stage in his hazy past.

Dinky, Max's pen-mate, said: "I hope he can cook better'n he can shear."

Max made the appropriate reply, and then asked if we could get the station cook over to prepare lunch and tea as he didn't want to spend the whole day in the kitchen. I didn't want him to, either.

This was the last shed of the run and we were trying to finish it by the end of the following week so we could all go home without having to spend another weekend in the bush.

We were in a good mob of ewes and Max had been shearing exactly 200 sheep a day for the past four days. We all knew he was going to make it a thousand for the week. That's a tremendous effort, in any kind of Australian sheep. But by pulling on the job of the departed cook, Max had obviously denied himself the magic target.

David offered to light the fires for Max and I left them to it while I drove to the station, about two kilometres away.

We couldn't have the cook, the owner said, but his own wife might be able to help. We trooped around to the small laundry shed at the rear of the homestead where his wife stood near the back wall, bending over an open copper tub set in a cement block and heated by a wood fire. Dry chopped wood was stacked against one side of the block. On the other, attached to the wall, were three roomy concrete rinsing troughs, each with their own cold water tap. A clothes wringer, hand-driven, was bolted to the side of the trough nearest the copper tub.

In the simmering, sudsy tub was a load of washing which the owner's wife was stirring with a bleached wooden rod. We popped the question.

"Yes, of course. I can fit that in. I'll just finish a couple of loads of washing and I'll be over. About an hour or so. The rest of the washing can wait till tomorrow, or the day after."

Her creased face wrinkled into a grin as she added, "I don't think it will rain."

She was one of those unflappable bush women who can fit almost anything into their day without making a fuss. Cook some scones. Skin a sheep. Fight a bushfire. Shoot a dingo. Have a child. Change a tyre. Whatever had to be done while the men were out working on the station. Suddenly being called on to cook a couple of meals for eight hungry men didn't cause a ripple.

She'd be there to make lunch, afternoon smoko and tea. With that under control I wheeled back to the quarters for breakfast.

Mowie was loose when I got back. One of the shearers had slipped his chain and he was enjoying an early morning romp.

Max served a decent breakfast of liver, bacon and eggs, and as we left him to start work in the shed, he promised us a batch of the lightest fruit scones we'd ever eaten.

"I've got to put the fruit in 'em," he said. "It's only the extra weight of the sultanas that keeps 'em from floating off the plate!"

After the doughy cannonballs served up by Cec, anything would have been an improvement.

In the shed Herbert saw his chance to hit the front. He was the shearer holding number two position behind Max.

Holding down the gun position is a matter of pride to many shearers, and Herbert coveted it. He'd been trailing Max by two or three sheep a day, and try as he might he couldn't close the gap.

They were both champions and it was a pleasure to see them labouring — "poetry in motion" someone called it.

Now with Max out of the way for a whole run, Herbert had the chance to make up a lot of lost ground.

The New Zealander rattled through fifty-one sheep before the bell sounded to end the first run. Five minutes before the bell, Max appeared in the doorway, silhouetted against the bright light outside and carrying the urn of scalding tea and a basket of scones, safely weighed down with sultanas.

He dumped the urn and basket on an overturned bale near the scales and moved to the end of the board to watch Herbert finish his last two or three sheep for the run.

The scones were indeed an improvement and we debated the possibility of keeping Max in the kitchen, but the station owner's wife had just arrived at the quarters, and with smoko out of the way, Max returned to his stand to prepare for the second run. Herbert looked set to ram home his advantage. He might hit the front yet. I rang the bell.

Max's sprint across the board to catch his first sheep would have done credit to a greyhound leaving the trap. He was off and running!

Good shearers define themselves by being able to increase speed without a proportional decline in workmanship. Max was such a man. He worked at a furious pace but he retained his easy, fluid style. All his shorn sheep were beautifully uniform, with very few skin cuts. Even more surprisingly, despite his constant threats to struggling sheep — "If yer don't keep that bloody leg still I'll snap it off and shove it down yer bloody neck" — I never saw him hit one.

Usually, by the time he'd finished his gripe, he'd finished the sheep, and that might have had something to do with it. Whatever the reason, he was a top shearer.

When he strode, sweat-soaked, from the board at the end of the run he threw me a wink from beneath a dripping eyebrow. It didn't take me long to discover why.

Sixty-seven sheep were in his let-out pen! He'd gone up seventeen a run.

Herbert's face was a picture when he picked up the tally book after lunch. He'd known that Max had shorn more than him but sixty-seven floored him! Herbert himself had dropped back to forty-nine.

"Struth," he said, "at that rate he'll still get his two 'undred!"

That's exactly what happened. In six hours shearing, instead of the full day's eight, Max shore 200 sheep to reach his thousand for the week.

Dinky gave Max some stick in the mess that night. "'Ere, Max, you've been bludging on us all this time. If yer'd been pulling yer weight all along we'd be 'ome by now."

"I'll have to lift me game then, Dink!" Max answered.

CITY LIGHTS

LATE AFTERNOON, just a few kilometres out of Sandstone. A silent shed. Empty yards. Quarters almost deserted, the mess door swinging and creaking in the breeze. A few final remarks on the wool report. Fold it up and stuff it in an envelope. Hand a copy to the owner. Shake hands.

"See you next year."

"Yep."

Eric finished packing the leftover stores into my van. He and I were the last to leave.

Apart from the pleasant interlude at Coral Bay and one quick trip to Perth, we'd spent much of the year in tiger country, so we were very pleased to be heading south for a rest. Mowie jumped into the middle of the front seat and we were ready.

Travelling made Eric thirsty.

We'd covered about five metres when he said, "Want a beer Hewsey?"

"Why not?" A couple of cans were enough for me. But not for Eric. He cracked another couple. And another couple after that.

"Go easy, mate!"

"No worries smate!"

Kilometres slipped by. Eric drained the Esky as we drove into the night. Then dusty yellow lights appeared ahead.

"Paynsh Find, Hewshey."

"Time for a break."

We eased to a stop in front of the Payne's Find hotel. The cramped little pub was always a welcome stop. I rubbed my eyes and wandered into the cool bar for a squash. Mo wandered round squirting every bush and post in sight.

Eric had a few beers and bought a few for the road.

"Crikey Eric. We'll never get to Perth. We'll be stopping every ten minutes for you to have a leak!"

"No worries smate. Sobers sjudge!"

"You're drunk!"

"Bullshit! Bull . . . shit! Sobers sjudge."

At that point, standing upright was a bit too much for Eric so he leaned against the fender while he rummaged for his cigarettes and matches.

Then he tried to light a cigarette. Well, he tried to get it into his mouth first. That wasn't very successful for a while but he eventually managed it.

Then he tried to strike a match. That fixed him. He either missed the side of the matchbox or hit it so hard that the match snapped. When he did get it right and it burst into flame he couldn't pinpoint the end of the cigarette before the match burned down to his fingers.

"You're pissed!"

"Sobers sjudge! Jus stired."

"Have a sleep then."

"Jus sfor a few minutes sthen I'll give you a spell drivin'!"

"Righto. Get in."

Mo jumped in. Eric fell in.

He was asleep before I could get his seatbelt fastened. And he stayed that way all the way home, well most of it anyway. After an hour or so he stirred.

"Jee Schrist I'm crook."

Then he straightened himself up and looked around in fearful panic. When I switched the cablight on I could see Eric's face doing a reasonable imitation of a bugler blowing a high note. Beaded brow. Bulging cheeks. Bulging eyes. Only the bugle was missing.

Clouds of rubber smoke hung in the cool night air as we skidded to a stop. Eric got his door open, just.

"BLLLUUUUUUUURRRRRRRRRRCHHHHH!! HUURRK! HUUUUURRRK! HAAAAWWWWKK! HOOOORRRRR!"

A torrent of cold mutton, diced carrots, peas and beer gushed out on to the road.

"Sbetter." Just a whisper.

"Glad to hear it!"

Back on the road. Eric went back to sleep.

After that we drove non-stop. Through mulga-studded pastoral country, then through the wheatbelt. At last we reached the first signs of a softer landscape . . . the Swan Valley.

Grey dawn cracked the night sky as we slipped through Midland. Perth was stirring from slumber as we sped through the outer suburbs towards Innaloo.

About six hundred kilometres had rumbled under the van's wheels and I was dog-tired. On the other hand, Mowie had slept the whole time and looked bright and alert. Eric was slumped and snoring.

A couple of strange vehicles were parked on the verge outside Peter's place; no doubt there would be a couple of strange bodies inside — but there were plenty of beds.

The back door was open and Eric let himself in, found a bed and collapsed again. I fed Mowie, showered, and tumbled thankfully in to another cleanish bed. We slept into early afternoon.

When sleep retreated, we discovered that Peter had left the day before for a spell in the eastern states. Apparently he'd met some fancy woman and they'd gone off to have a holiday together.

My first priority was clothing. After eight months in the scrub, my wardrobe looked like something discarded from the Burke and Wills expedition.

A men's store in the Dianella shopping centre seemed promising, but Mowie prevented me from going in. With that elegance which only dogs possess, he hunched himself up, stiffened his tail and grunted his way through the deposition of a nice mess on the shop's "Welcome" mat. He went through the ritual of raking the mat with his feet but, by that time, I was half the carpark away, heading for the van.

At the next shop he stayed in the van, and just before closing time I emerged with bundles of new clothes. Now for a new car. It was Wednesday night, late-night trading for car dealers in Perth.

Brightly lit yards lined the Albany Highway all the way from Perth through Victoria Park and on to Cannington.

Jim Lloyd worked at a yard in Cannington. "Struth," he said, "it's bloody Hewsey! How the hell are ya!"

Then his office phone rang and he greeted the caller with a polished: "Good evening. Jim Lloyd heyah, how can I help you?"

Apparently the voice on the phone wanted to buy a car. Jim made an appointment to see the caller and hung up with a silvery: "Thank you very much, very much indeed, goodbye!"

I wondered if I would get the same treatment.

The next moment he jumped up and squeezed my outstretched hand. "Good to see ya again, Hewsey. How long y'down for?"

When I told him six weeks, he thought he might be able to get in a couple of days' fishing in his boat.

When I said I'd also come to buy a car, he really did look pleased to see me.

"What do you have in mind, Wade?" he asked, almost slipping into his polished mode, but laughing out loud just before I was able to comment on it. "What are you chasing — another van?"

He showed me a V8 with bucket seats. "This one's only a couple of years old and I can do you a real good deal on it. Take the thing out for a burn. See what you think . . . er . . . don't get any dog hairs on the upholstery, will you . . . ? Not until you've bought it anyway."

I ordered Mo off the seat and into the back. Albany Highway was choked with traffic, but a gap appeared and I rammed the accelerator down. V8 power swished us into the traffic and south towards William Street. Down to the railway. Along Sevenoaks Street. Back to the Highway. Turn left. Back to the yard. It certainly seemed to drive well, but after the rattles and draughts of the old van anything would have been an improvement.

Sold.

But hopes of a good trade-in figure received a setback when the car yard valuer came to put a price on my old van. As soon as he laid eyes on it he blurted out: "Good God almighty! Was anybody seriously hurt?" He then inspected every centimetre of it, every dent and scratch.

Jim was apologetic when he told me what the valuer thought it was worth: $800.

"Is that all?"

"Yup."

The thought of my faithful old van, now standing forlorn and abandoned in the trade-in yard, about to finish its days in a wrecker's yard, was quite sad, but I signed up.

"The new van will be ready for delivery tomorrow afternoon, Hewsey."

When I pulled up at the yard next day, Jim took me over to my new acquisition. It had been cleaned and polished. All the paperwork was ready to be signed.

"And I've filled the tank for ya, Hewsey." He shook my hand and opened the driver's side door for me to get in. "Thanks for the business, mate. Don't forget Satd'y. We'll go fishing."

☆ ☆ ☆

Saturday was only five hours old when we met at the East Street ramp in Fremantle. The ramp faces a lovely stretch of the Swan River, but it's only a short run through Fremantle Harbour to the open sea.

Jim was proud of his boat. He undid the restraining ropes, lined the trailer up and sped off in reverse down the ramp. As soon as the trailer wheels hit the water, he jammed on the brakes and the sleek red hull slipped graciously into the river.

Jim then zoomed off to park the car, leaving me up to my thighs in cold water holding his boat. He hadn't gone for comfort when he'd bought it. It was an open 16-footer, fibreglass, with a 50 hp outboard motor. And it was sinking!

Frantically heaving and yelling, I managed to drag the stricken craft into shallower water just as Jim came sprinting back from the car park.

As the boat settled to its gunwhales, he commented morbidly: "Blast! It's a bastard if you forget to check the bungs."

Within an hour we'd bailed all the water out and refloated the boat. Finally we set off downriver for the open sea. A three-can trip down to the heads at eight knots. Then Jim opened the throttle and we zoomed out into the ocean. He was heading for a couple of large piles that stuck out of the water a few kilometres out from Fremantle.

"Drop the pick here, Hewsey, just by this first pile. No wait . . . we'll go a bit further . . . drop it by that second pile . . . now!"

The anchor went over with a splash. Without realising it, Jim had just positioned us right in the middle of the shipping channel.

As the morning wore on, we caught a few herrings and garfish and drank a few more cans. The water was like a mirror. Sunbeams glanced off the surface and sparkled all around. Warmth from the sun made us sleepy. It was so peaceful. Only the distant buzzing of other motor boats disturbed the quiet.

Then we became aware of a persistent *Hoooooooooooooot! Hooooooooot! Hoooooooooot!* somewhere in the distance behind us.

We took no notice. Until it grew very loud. At which point Jim glanced back over his shoulder.

"Bloody hell! Look out! Get the anchor! Get the anchor!" He sprang for the outboard motor and furiously pulled the starting rope.

When I looked back, I nearly flipped. The huge black bow of an oil tanker was looming and almost over the top of us. *Hooooooooooooot! Hooooooooooooot!* An evil-swelling bow-wave was foaming towards us. A bloke with a megaphone was shouting at us from the bridge. I had the anchor just clear of the bottom when the motor fired. Jim didn't wait. Into gear. Full-pelt forwards. The sleek, red hull speared away. I fell backwards, flat out on the floor of the boat. I had an upside-down view as the tanker slid past, with the bloke on the bridge hurling abuse and shaking his fist at us.

"Struth!" said Jim. "Did ya see that!"

That was enough fishing for one day.

8
A Cool Change

SCANNING *The West Australian* revealed quite a few sheds starting reasonably close to Perth. Down south. And the southwest of Western Australia was a refreshing change from the arid north. Green paddocks. Trees. Blue sky and fluffy white clouds.

I picked up three small sheds to do while we waited to restart up north. It was as good as a holiday.

The pace in these smaller southern sheds was much more relaxed. It was cooler. The environment was more peaceful. And we were close enough to Perth to spend every weekend there.

Some of the team were locals who even went home in the evenings. That meant that we weren't always living in each other's pockets, and there was much less friction in the team.

The last shed in this short run belonged to Fred Evans. Fred loved his sheep. He could rattle off the life history of most of his small flock. That one had been born under the old harvester in the corner of the bottom paddock. This was one of twins. The other had died and the survivor had looked pretty sick for a while but it had come good.

Fred and his wife owned a choice block. It carried only about three thousand sheep, but it carried them well. Feed was naturally abundant, and on the infrequent occasions when it dwindled, Fred would buy in the best hay available and hand-feed his stock. His compassionate methods paid handsome dividends. The new crop of lambs looked healthy and robust. The ewes were radiant and the wethers were huge.

In fact, Fred told me that one team of shearers decided the

wethers were too big and refused to start work on them.

"I'd have offered them more money," he said, "if they'd only asked for it. But they didn't even do that. They just drove in, saw the size of the wether hanging up in the meathouse, turned round and drove out again. Never even stopped to say 'hello'. Took me a week to find another team."

Fred had learned from that experience. There was a plump lamb hanging in the meathouse when we arrived. It was eaten in a couple of days — and now it was time to kill a wether.

Half a dozen wethers had been put through the shed with the ewes; I'd drafted them off when I counted the sheep out of the shearers' let-out pens. There'd been some rumblings from the shearers when they'd tackled these giants. "Phar Lap with a fleece" was how one of the blokes described them.

Now, I discovered that Fred didn't like killing the sheep himself. There were two reasons. To begin with, he had a stiff leg. One of his enormous rams had butted his knee against a post in the yards, as a result of which Fred was hospitalised and never bent his leg again.

Slaughtering sheep demands some agility. So he had to call on neighbouring farmers to kill his meat for him. Secondly, Fred was too attached to his flock to wield the knife himself.

On this occasion I'd offered to do the dirty work. We stood together looking into the pen of killers, deciding which one it should be.

"That one looks just right, Fred."

He agreed and I jumped the rail to catch the doomed animal. The killing block was about twenty metres from the yards.

With standard-sized wethers it's simple enough to straddle them, grab their horns or ears, guide them forward through the gate and then, however reluctantly, towards the block. But when I straddled this one, my feet were off the ground. Instead of guiding it anywhere, I became a passenger.

That wether and I gave a reasonable imitation of a bull ride at a country rodeo. There was no way I was going to get it out of the yards alive.

"I'll cut its throat here, Fred. Then we can drag it over to the block."

"Right."

Tipping the animal over took a supreme effort but eventually I had it down on its side, head pulled back, knife poised. The idea is to slash the throat with one stroke, then bend and twist the head back until the neck snaps. I struck with the knife but at the first bite of the sharp steel the wether grunted in panic, heaved itself to its feet and tipped me over.

The damage had been done. Blood gushed from a mortal wound but that didn't stop the terrified animal from dashing drunkenly round the yard, rolling its eyes and making piteous noises that can best be described as a cross between a *baaaa* and a gurgle.

Fred went white. He didn't look at all convinced when I told him: "It's dead, Fred, that's just its nerves twitching."

Eventually, in a last dying spasm, the blood-soaked wether lurched into a corner of the yard, rebounded back on to its haunches and collapsed. Its eyes went glassy and the great body quivered into stillness.

I turned to look at Fred but he was gone, his stiff leg carrying him back to the homestead as fast as it could go.

☆ ☆ ☆

Time passed too quickly in this soft, fertile corner of the State. We were to head back north and I really didn't want to go. But the time had come, so I rang Peter Margate to find out when we were due to start. Eric answered the phone.

"When are we off, Eric?"

"We might not be going!"

"What?"

"Pete's in the 'nure with the tax man and he's shot through."

"So where's Pete now, Eric?" I asked.

"Dunno. I'm going to try to take over. Do you want to come and give us a hand to run the show?"

"I'll let you know."

It was a tough decision. Here was a chance to become a contractor. What could that mean? It might mean a substantial income on top of my classing wages. It would certainly mean a lot of extra work and organising. I'd done enough of that already to know what that would be like.

And, to make a real go of it, it would mean at least several more years in the isolated northern pastoral areas.

When I phoned back, I had made up my mind. "That you, Eric?"

"Yep."

"I'll give it a miss, mate."

"Okay."

"Good luck."

"Yeah, thanks."

ON THE FRONTIER

UNEMPLOYMENT BECAME a companion again. But spring was turning to summer, so there would soon be plenty of work about.

Meanwhile, there was time for a quick trip to Adelaide. Mowie stayed at the Butlers.

Days at the beach. Nights out with my mates. It was a great couple of weeks. Then the phone rang. Mum answered it. "It's someone from Western Australia for you, Wade."

"Hello?"

"My name's Griffith. I'm a shearing contractor and I've got a run around Esperance. Do you want it?"

"How long for?"

"Ten months?"

"How much?"

"Award plus a bit."

"I'll take it — but how did you find me?"

"Geoff Gordon told me to give you a call. He shears for me. Said you won't take any stuffing around."

Struth! Geoff Gordon. The shearer I'd sacked for telling me to get stuffed.

"Tell Geoff thanks very much. When do you want me to start?"

"Can you be in Esperance on Mond'y?"

Today was Friday!

☆ ☆ ☆

Esperance was pioneer country. Big farming syndicates. Small cockies. Both trying to carve viable farms from the sandy soil.

Wind erosion was one of their biggest problems. When they cleared huge areas of native mallee scrub, the loose, dry soil just blew away. Clouds of dust were billowing up in all directions as I drove out of town and headed north towards Gibson Soak.

Just past the pub at the Soak, a track wound off to the east. Down that track was the first shed of the run.

The cocky was typical of the smaller landholders. He was struggling to make a go of it. Most of the farms in the area were mixed: they grew crops and grazed sheep, cattle and sometimes pigs. Anything that would return some money.

Establishing a new farm, especially one in such a remote area, was an enormously expensive job. Fences, machinery, sheds, water tanks, chemicals — almost everything had to be carted in by long-distance transport. No wonder, then, that most of the sheds and living quarters were ramshackle affairs.

This first cocky told me that his farm's immediate future depended on this woolclip. Money from the sale of wool would clear some of his debts and give him some capital to invest in fertilizer and fuel to put his next crop in.

He was a gruff but friendly bloke with battered hands and black, split nails.

"Last year's classer ran around like a chook with his 'ead cut orf," he said. "'E 'ad 'eaps of wool everywhere!" Inside the shed it quickly became obvious why.

To begin with, there were only four wool bins. Then I looked at the previous year's list of pressed bales. There were over a dozen types of fleece wool listed! On a small farm that can often be a sign of overclassing. The idea is to have as few types as possible. Most woolgrowers know this and aim for consistency among their sheep. Any that produce wool that's a long way off the average for the farm are regularly culled out. But the cockies at Esperance couldn't afford that luxury.

When the government threw open the Esperance area, sheep

were scarce. Small cockies had to scrounge their stock from wherever they could find it. A few hundred sheep from up north, a few more from over the border in South Australia. A mob or two from Queensland. Even some tottery old ewes saved from the butcher's knife. The idea was to get sheep on to the land. What kind of sheep was not important.

So, in the pens and holding yards of this shed there were pure Australian merinos and long-wooled English breeds, and every possible combination of them both. They stood there, looking back at me and chewing their cud. I chewed my bottom lip.

"What do you reckon, then?"

"I reckon I'll be running round like a chook with its head cut off, too!"

I made one attempt to pre-class the sheep by marking the crossbreds with wide, blue-raddle stripes. I asked the shearers to shear those last. It worked for the first pen, but by the time the shearers called for more sheep I had so much wool lying around in the dingy woolroom, I couldn't afford the time to raddle the next lot. So after that I just classed the fleeces as they came.

When I compared my list of wool types with the last year's, I was convinced that the previous classer had done a good job.

Most of Esperance sheds were like that. It was sheer hard work.

Pleasant spring warmth ignited into the sizzling heat of summer and I began to question seriously whether this was the life I really wanted to lead.

There was certainly an air of the frontier about the place, but it wasn't much fun. We were living and working in the roughest conditions. The wool was almost always full of heavy sand, which blunted the shearers' combs and cutters and made the wool fall to pieces when it landed on the table.

The air was full of dust and we started each run by smearing a thick wad of petroleum jelly up our nostrils. That helped to trap the dust and made it easier and smoother to dig out with a grimy forefinger.

We were over six hundred kilometres from Perth but I started going back every weekend. That caused a few problems for George Griffith. Spurred by eagerness to reach Perth, I regularly drove the distance in under seven hours. There was always someone else in the team who wanted to go to Perth for the weekend.

Most of them were willing enough to travel with me once. Some even came with me twice. But after a particularly swift journey to the city, a couple of them refused to get back in the van for the return trip. George was very cross. He only found out when I rolled up on Monday morning, and he had a mad scramble to find a couple of replacements.

After that I travelled to and from Perth alone every weekend.

George had his own small property near Esperance. A few sheep, some cattle, about a hundred pigs. He'd sell a plump baconer to one of his shearing team for ten dollars. "But you've got to knock it off and butcher it yerself," he said.

I'd never slaughtered a pig before, but I paid my tenner and picked up the stunning mallet. According to the time-honoured technique, you knock out the pig with a clean, well-aimed blow on top of the head, then slit its throat and bleed it.

This pig didn't want to die. It dodged the first blow and tears came to my eye when the whistling mallet arced down and smacked into my shin. Several glancing blows followed, all of which produced little more than horrifying squeals from the pig. Squeals echoed a hundredfold from the other panic-stricken pigs stampeding round the yard next to the killing pen. Eventually I cornered the poor creature and got a solid thump in between its ears. It squealed and went down.

At last. I grabbed one ear to jerk its head back and reached for the knife. The pig revived, jumped to its feet and bolted. More squeals. More furious swings with a bloody mallet. Then a heart-rending shrill scream. A death scream. The pig quivered into stiffness and plunged to the ground. I went down on top of it, heaving and sweating. The pig's eyes closed. Mine bulged like a psychopathic murderer's after a massacre.

That was the easy part over! Now I had to get all the bristles off. For that purpose George had a trough made from half a 44-gallon drum. It was filled with fresh water. A fire was lit beneath it and when the water reached the right temperature, the dead pig was dropped in.

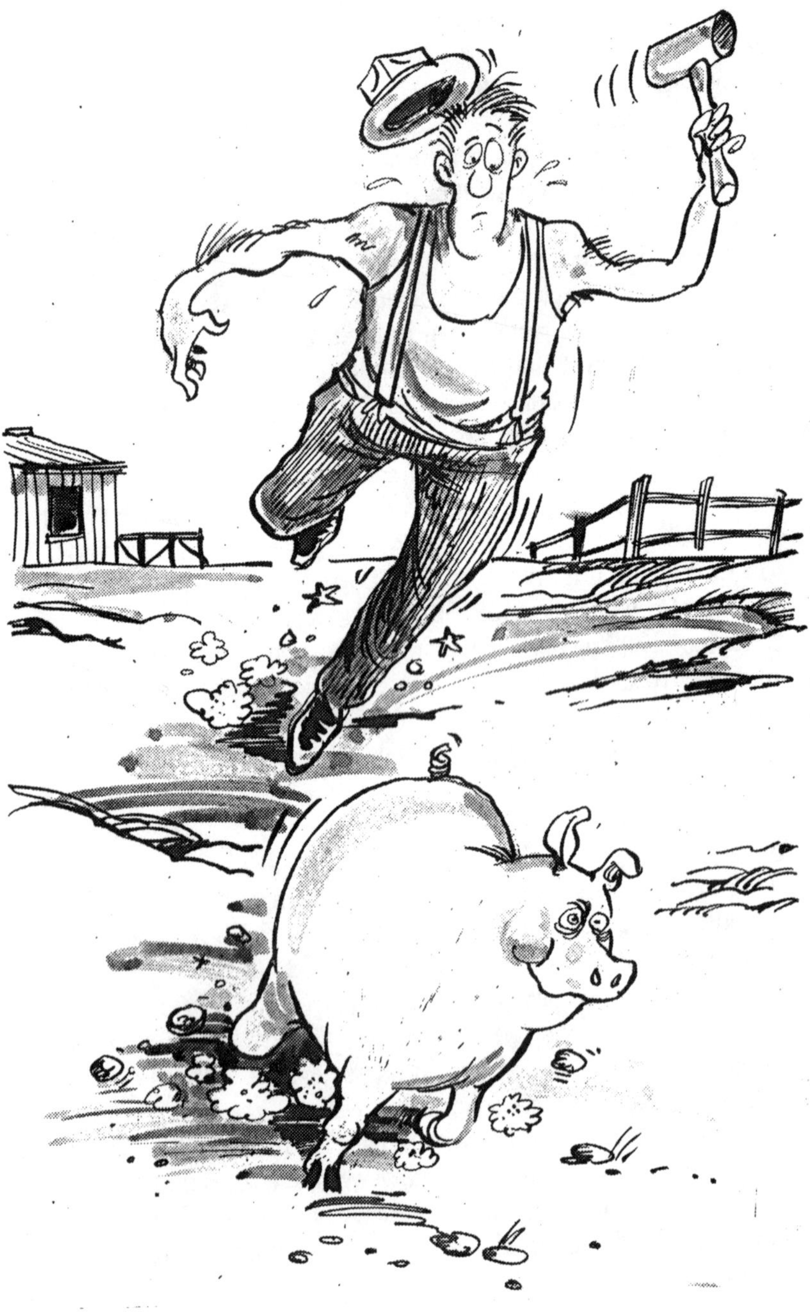

In theory, this loosened the bristles so that they could be easily scraped off with a large fish-scaler.

The theory didn't seem to work so well for me.

After a couple of hours scraping and plucking, I had most of the bristles off the body but the head was still quite hairy. I gave up and hung the carcase in the meat-house.

Next morning was Friday, so I got up early, cut the pig's head off, then carved up the carcase into halves, then quarters. I wrapped the head and meat in muslin bags and hung them up again.

That night, they were in the back of the van heading for Perth. The Butlers would be eating pork for a month.

Mrs B was most enthusiastic about the head. "That'll make good soup," she declared.

"I'll have to get the bristles off for you first, Mrs B — probably one at a time," I told her. "If you've got a pair of tweezers, I'll do it first thing in the morning."

Off to bed I went, dog-tired from the long drive.

When I shuffled, bleary-eyed, from my room on Saturday morning, Mrs B was already busy. She was sitting at the kitchen table. In front of her was the pig's head, enjoying an early morning shave with Mr B's safety razor!

UP TO MY KNEES IN IT

THE WORK AT Esperance ground on. Rough sheep, rough conditions, good blokes, but little enjoyment. Daydreams of better times closer to home.

Then two things happened to make me decide to give up the bush.

The first occurred early one Friday morning. The night had been stifling. Mozzies had feasted on us all night as four of us slept in a decaying plywood caravan. The cocky had towed it out to his block years ago. He'd lived in it while he built his small house. Now a peeling, rough-daubed sign nailed above the door announced it was the shearers' quarters. Inside was a mess. The walls were curling away from the frame. The roof had holes, and dead weeds protruded through cracks between the floor and walls.

I had one of the top bunks. It was six-thirty when I swung my legs over the edge and dropped to the floor. Well, through it, in fact! The plywood was rotten and gave way underneath me. I ended up standing on the ground, up to my knees in caravan, with both my shins scraped raw.

Nobody laughed.

A foul mood stayed with me all day. Months of these conditions had taken their toll.

That night, I swung the van out of the gate and roared up the road towards Perth, still feeling morose. About eight o'clock, with the speedo showing something over 160 kmh, I ran off the road on a corner. The van contributed to the land-clearing programme by scything down a sizable area of wiry scrub before it finally stopped. The noise was awful. We were still upright though, and apart from one of the best frights I've ever had, I was okay.

Mowie eloquently expressed his feelings by bolting out of the van and into the bush as soon as I pushed the door open against the press of leaves and branches.

Three hours later I'd shifted enough scrub and sand to be able to drive the van backwards on to the bitumen again. Getting Mowie to come back and get in the van consumed another twenty minutes.

The roo bar was hard up against the bonnet and the steering wobbled a bit, but everything else seemed all right. We turned and headed back towards Esperance.

Next morning I drove the van to the local garage for a check-up and repairs, then rang George.

"I've had enough, mate."

"When do you want to leave?"

"As soon as we finish this shed."

"Okay. Do you want to sell yer dog?"

"No."

☆ ☆ ☆

Smooth bitumen glided under the wheels, suppressing at least some of the rattles in my quickly aged van. Mowie snored on the seat in his usual position, his head on my lap. Patches of mist, lying in the hollows, slipped by, silently reflecting the headlights back into the cab.

The radio, tuned to a Perth station, gradually gained strength as we drew closer to the bright lights. Another few hours and the bush would be behind us for ever.

A time for reflection? Not yet. I was too glad to be free of the rigours of bush life, too glad to be going back, after seven years, to a normal existence. I was looking forward to not having to live out of a suitcase, not having to put up with hot, sleepless nights, followed by hot, gut-busting work.

Instead there'd be a normal social life. Evenings at home. No need to get out of bed early on Sundays to be sure of getting a trough to do my washing. No need to slaughter sheep for the team's meat supply every other night.

In fact, no more sheep!

I didn't know what the future held. Didn't know what sort of a job I'd get. But I did know it wouldn't be anything to do with sheep.

Just then a blur appeared in the edge of the headlights and swerved directly into our path. A crashing thud jarred the van and in a moment we were slewing round as the vehicle ground to a stop. Jammed under one of the back wheels was the body of the victim — a ewe.

Pitifully bleating by the side of the road was her new-born lamb, saved from the same gory fate only because wobbly legs had been unable to keep up on that suicidal dash.

Left alone on such a cold night, it probably wouldn't survive long enough to be killed by foxes. It didn't even try to escape when I stooped to pick it up. I tossed up between knocking it on the head or giving it a lift to the nearest homestead. Poor sod. Its terrified heart was beating a rapid tattoo inside its skinny chest. Small, faint clouds of condensation puffed from its pink muzzle with every breath. Bigger clouds with every desperate bleat. I couldn't bring myself to kill it. The only way to keep it from thrashing around in the back of the van was to wrap it tightly in a blanket.

Mowie sat on the front seat, staring back at the swaddled orphan as we continued our journey.

Dawn had just ripped the night sky when a homestead appeared off to the right. A yellow window-light signified someone was up, so I swung into the driveway. Wood-smoke, blue and pungent, curled from the chimney.

The farm was on the very fringe of the wheatbelt. Perched precariously between profitability and poverty. It all depended on the weather. In good seasons the semi-desert climate retreats a little. Rain comes and the marginal lands grow crops and stockfeed. In poor seasons the bone-dry sky sucks life from the land and life savings from the bank.

The farmer's wife answered the door before I had the chance to knock. She looked as though she'd seen too many droughts. Forty-ish, a faded woman in a faded cotton dress and misshapen woollen cardigan. When she saw the lamb her face lifted in a weary smile.

"Yes. I'll look after it. This'll be three pet sheep I've got. Had one of them for four years. Reg is always threatening to send them off to the meatworks, but he wouldn't do that. Would you like a cup of tea?"

"Thanks."

"Bring the lamb in with you."

Threadbare lino covered the kitchen floor. An offcut of the same lino covered the kitchen table. It was cross-hatched with countless scratches. Fresh breakfast dishes were stacked on the stainless steel sink under the window.

"Reg left early this morning to work on one of the windmills. It hasn't been pumping for a couple of days and we had to wait for some parts to come up from Perth."

I sat on a green, straightbacked chair opposite the window. The lamb settled comfortably on my lap, sucking my thumb as the woman rummaged in a kitchen cupboard. Eventually she found an old baby's bottle. She mixed dried milk powder with warm water in a cracked pottery jug, added a teaspoon of honey, filled the baby's bottle and handed it to me.

While I tried to get the lamb to drink, she got a pot brewing. By the time my enamel mug of tea was poured, the lamb was guzzling greedily from the bottle. Droplets of dribbled milk gathered on the silky hair around its lips.

I let it finish the bottle before I reached for the tea. The woman took the lamb from my lap and disappeared through a side door with it tucked under one arm. Minutes later I could hear it bleating somewhere outside.

The woman reappeared through the door.

"I've got it in our little shed. It'll be safe and warm there. I call it the nursery." She laughed.

A sudden and totally unexpected feeling of guilt swept through me. I was able to leave this environment; this woman and her husband were chained to their grudging farm. As an itinerant woolclasser, I'd been a spectator of life in the bush. These people, and many others like them, were living it.

They'd probably spend their whole lives struggling against the elements and the mortgage, unable to influence the price they obtain for their produce and forced to pay the going rate for raw materials and spare parts.

Time to go. I swung the van back on to the main road as the sun was daubing its first rays of warmth over the ground. The straggly bush and the pockets of grim farmland seemed to tense themselves for another hard day. The bloody bush. It was time to get out of it. The humour had gone.

Epilogue

ONE AUGUST DAY, one of those gloriously sunny days when you wouldn't be dead for quids, my phone rang. On the line was Mr Butler, virtually my foster-father from the time I'd arrived in Western Australia.

The Butlers' house had been my home for eighteen months, and when I'd moved out to live in a flat, Mowie had stayed on with the family. His incorrigible nature had won him a permanent home there, and he'd settled down well. I enjoyed visiting rights, but Mowie became the Butlers' dog. They loved him.

"Just like him," I'd often thought, "to end up with a whole family of butlers at his beck and call."

But he was always glad to see me and he was my only tangible link with our bushy past. Now Mr Butler had called to tell me that a vet had administered the needle which rang the cut-out bell for Mowie. He'd had lung cancer.

If Mo had been able to read this book, I hope the little black-and-white bloke would have considered it to be a fitting memorial for him.

Photo by WAYNE OSBORN

WADE HUGHES and his wife Robyn live in Portland, Victoria. There, surrounded by rolling pastures, forests and the sea, he writes in his spare time.

There'll Never Be Another Ewe is his eighth book.